160
Advances in Polymer Science

Editorial Board:
A. Abe · A.-C. Albertsson · H.-J. Cantow
K. Dušek · S. Edwards · H. Höcker
J. F. Joanny · H.-H. Kausch · K.-S. Lee
L. Monnerie · S. I. Stupp · U. W. Suter
G. Wegner · R. J. Young

Springer
*Berlin
Heidelberg
New York
Barcelona
Hong Kong
London
Milan
Paris
Tokyo*

Filled Elastomers
Drug Delivery Systems

With contributions by
M. Arora, G. Carlesso, J.M. Davidson,
A.J. Domb, G. Heinrich, M. Klüppel, E. Kozlov,
M.N.V. Ravi Kumar, A. Prokop

This series presents critical reviews of the present and future trends in polymer and biopolymer science including chemistry, physical chemistry, physics and materials science. It is addressed to all scientists at universities and in industry who wish to keep abreast of advances in the topics covered.

As a rule, contributions are specially commissioned. The editors and publishers will, however, always be pleased to receive suggestions and supplementary information. Papers are accepted for „Advances in Polymer Science" in English.

In references Advances in Polymer Science is abbreviated Adv Polym Sci and is cited as a journal.

Springer APS home page: http://link.springer.de/series/aps/ or
http://link.springer-ny.com/series/aps/
Springer-Verlag home page: http://www.springer.de

ISSN 0065-3195
ISBN 3-540-43052-0
Springer-Verlag Berlin Heidelberg New York

Library of Congress Catalog Card Number 61642

This work is subject to copyright. All rights are reserved, whether the whole or part of the material is concerned, specifically the rights of translation, reprinting, re-use of illustrations, recitation, broadcasting, reproduction on microfilms or in other ways, and storage in data banks. Duplication of this publication or parts thereof is only permitted under the provisions of the German Copyright Law of September 9, 1965, in its current version, and permission for use must always be obtained from Springer-Verlag. Violations are liable for prosecution under the German Copyright Law.

Springer-Verlag Berlin Heidelberg New York
a member of BertelsmannSpringer Science+Business Media GmbH
http://www.springer.de

© Springer-Verlag Berlin Heidelberg 2002
Printed in Germany

The use of registered names, trademarks, etc. in this publication does not imply, even in the absence of a specific statement, that such names are exempt from the relevant protective laws and regulations and therefore free for general use.

Typesetting: Data conversion by MEDIO, Berlin
Cover: MEDIO, Berlin
Printed on acid-free paper   SPIN: 10753265   02/3020hu - 5 4 3 2 1 0

Volume Editor

Prof. Kwang-Sup Lee
Dept. of Polymer Science & Engineering
Hannam University
133 Ojung-Dong
Taejon 306-791 , Korea
E-mail: kslee@mail.hannam.ac.kr

Editorial Board

Prof. Akihiro Abe
Department of Industrial Chemistry
Tokyo Institute of Polytechnics
1583 Iiyama, Atsugi-shi 243-02, Japan
E-mail: aabe@chem.t-kougei.ac.jp

Prof. Ann-Christine Albertsson
Department of Polymer Technology
The Royal Institute of Technolgy
S-10044 Stockholm, Sweden
E-mail: aila@polymer.kth.se

Prof. Hans-Joachim Cantow
Freiburger Materialforschungszentrum
Stefan Meier-Str. 21
79104 Freiburg i. Br., Germany
E-mail: cantow@fmf.uni-freiburg.de

Prof. Karel Dušek
Institute of Macromolecular Chemistry, Czech
Academy of Sciences of the Czech Republic
Heyrovský Sq. 2
16206 Prague 6, Czech Republic
E-mail: dusek@imc.cas.cz

Prof. Sam Edwards
Department of Physics
Cavendish Laboratory
University of Cambridge
Madingley Road
Cambridge CB3 OHE, UK
E-mail: sfe11@phy.cam.ac.uk

Prof. Hartwig Höcker
Lehrstuhl für Textilchemie
und Makromolekulare Chemie
RWTH Aachen
Veltmanplatz 8
52062 Aachen, Germany
E-mail: hoecker@dwi.rwth-aachen.de

Prof. Jean-François Joanny
Institute Charles Sadron
6, rue Boussingault
F-67083 Strasbourg Cedex, France
E-mail: joanny@europe.u-strasbg.fr

Prof. Hans-Henning Kausch
c/o IGC I, Lab. of Polyelectrolytes
and Biomacromolecules
EPFL-Ecublens
CH-1015 Lausanne, Switzerland
E-mail: kausch.cully@bluewin.ch

Prof. Kwang-Sup Lee
Department of Polymer Science & Engineering
Hannam University
133 Ojung-Dong
Teajon 300-791, Korea
E-mail: kslee@mail.hannam.ac.kr

Prof. Lucien Monnerie
École Supérieure de Physique et de Chimie
Industrielles
Laboratoire de Physico-Chimie
Structurale et Macromoléculaire
10, rue Vauquelin
75231 Paris Cedex 05, France
E-mail: lucien.monnerie@espci.fr

Prof. Samuel I. Stupp
Department of Measurement Materials Science
and Engineering
Northwestern University
2225 North Campus Drive
Evanston, IL 60208-3113, USA
E-mail: s-stupp@nwu.edu

Prof. Ulrich W. Suter
Department of Materials
Institute of Polymers
ETZ,CNB E92
CH-8092 Zürich, Switzerland
E-mail: suter@ifp.mat.ethz.ch

Prof. Gerhard Wegner
Max-Planck-Institut für Polymerforschung
Ackermannweg 10
Postfach 3148
55128 Mainz, Germany
E-mail: wegner@mpip-mainz.mpg.de

Prof. Robert J. Young
Manchester Materials Science Centre
University of Manchester and UMIST
Grosvenor Street
Manchester M1 7HS, UK
E-mail: robert.young@umist.ac.uk

Advances in Polymer Science
Now Also Available Electronically

For all customers with a standing order for Advances in Polymer Science we offer the electronic form via LINK free of charge. Please contact your librarian who can receive a password for free access to the full articles. By registration at:

http://link.springer.de/series/aps/reg_form.htm

If you do not have a standing order you can nevertheless browse through the table of contents of the volumes and the abstracts of each article at:

http://link.springer.de/series/aps/
http://link.springer-ny.com/series/aps/

There you will find also information about the

- Editorial Board
- Aims and Scope
- Instructions for Authors

Contents

Recent Advances in the Theory of Filler Networking in Elastomers
G. Heinrich, M. Klüppel 1

Pharmaceutical Polymeric Controlled Drug Delivery Systems
M. N. V. Ravi Kumar, N. Kumar, A. J. Domb, M. Arora 45

Hydrogel-Based Colloidal Polymeric System for Protein and Drug Delivery: Physical and Chemical Characterization, Permeability Control and Applications
A. Prokop, E. Kozlov, G. Carlesso, J. M. Davidson 119

Author Index Volumes 101–160 175

Subject Index 189

Recent Advances in the Theory of Filler Networking in Elastomers

Gert Heinrich[1], Manfred Klüppel[2]

[1] Continental AG, Strategic Technology, P.O. Box 169, 30001 Hannover, Germany
E-mail: *gert.heinrich@conti.de*
[2] Deutsches Institut für Kautschuktechnologie e. V., Eupener Strasse 33, 30519 Hannover, Germany
E-mail: *manfred.klueppel@DIKautschuk.de*

Abstract. The viscoelastic properties of (mostly carbon black) filled elastomers are reviewed with emphasis on the strain-dependence of the complex dynamic modulus (Payne effect). Considerable progress has been made in the past in relating the typical dynamical behavior at low strain amplitudes to a cyclic breakdown and reagglomeration of physical filler-filler bonds in typical clusters of varying size, including the infinite filler network. Common features between the phenomenological agglomeration/deagglomeration Kraus approach and very recent semi-microscopical networking approaches (two aggregate VTG model, links-nodes-blobs model, kinetic cluster-cluster aggregation) are discussed. All semi-microscopical models contain the assumption of geometrical arrangements of sub-units (aggregates) in particular filler network structures, resulting for example from percolation or kinetic cluster-cluster aggregation. These concepts predict some features of the Payne effect that are independent of the specific types of filler. These features are in good agreement with experimental studies. For example, the shape exponent m of the storage modulus, G', drop with increasing deformation is determined by the structure of the cluster network. Another example is a scaling relation predicting a specific power law behavior of the elastic modulus as a function of the filler volume fraction. The exponent reflects the characteristic structure of the fractal filler clusters and of the corresponding filler network. The existing concepts of the filler network breakdown and reformation appear to be adequate in describing the deformation-dependence of dynamic mechanical properties of filled rubbers. The different approaches suggest in a common manner that there is a change of filler structure with increasing dynamic strain. However, in all cases additional assumptions are made about the accompanying energy dissipation process, imparting higher hysteresis to the filled rubber. This process may be slippage of entanglements (slip-links) in the transition layer between bound rubber layer and mobile rubber phase, and/or partially release of elastically 'dead' immobilized rubber trapped within the filler network or agglomerates.

The theoretical understanding of filled elastomers has been improved to the extent that now a connection can be made between the filler structures on larger length scales and the viscoelastic properties of rubbery materials.

Keywords. Elastomers, Filler networking, Fractal filler structures, Payne effect, Viscoelasticity

1	Introduction and general statements	2
2	The Kraus Model and related approaches	6
3	The VTG-model	12

4	The Network Junction (NJ)-model	15
5	The Links-Nodes-Blobs (L-N-B)-model	18
6	The Cluster-Cluster Aggregation (CCA)-model	21
7	Conclusions	27
References		29

1
Introduction and General Statements

Elastomers are well known for their ability to withstand large deformations. The dynamic mechanical properties of unfilled elastomers at small dynamic strain depend upon temperature and frequency. When a sinusoidal shear stress is imposed on a linear viscoelastic rubber, the strain will also alternate sinusoidally but will be out of phase, the strain lagging the stress. Strain, γ, and stress σ, can be written as

$$\gamma = \gamma_0 \cdot e^{i\omega t} \tag{1}$$

and

$$\sigma = \sigma_0 \cdot e^{i(\omega t + \delta)} \tag{2}$$

where γ_0 and σ_0 are the maximum amplitude of strain and stress, respectively; t is time, and δ is the phase angle between stress and strain. The stress: strain ratio at any instant in time defines the complex shear modulus

$$G^* = \frac{\sigma(t)}{\gamma(t)} = \frac{\sigma_0}{\gamma_0} e^{i\delta} \tag{3}$$

This quantity can be resolved into real (G') and imaginary (G'') components such that

$$G^* = G' + iG'' \tag{4}$$

and

$$G' = \frac{\sigma_0}{\gamma_0}\cos\delta, \quad G'' = \frac{\sigma_0}{\gamma_0}\sin\delta \tag{5}$$

G' is termed the storage modulus and is proportional to the maximum energy stored per cycle whilst G'' is termed the loss modulus. The loss factor $\tan\delta$ is defined by the ratio

$$\tan\delta = \frac{G''}{G'} \tag{6}$$

Note that the loss tangent is also a measure of the ratio of energy lost to energy stored in a cyclic deformation. The energy loss density during one cycle of strain, ΔE, is given by

$$\Delta E = \int \sigma d\gamma = \int_0^{2\pi\omega} \sigma \cdot \dot{\gamma} dt \tag{7}$$

From Eqs. (1)–(5), one has

$$\Delta E = \pi \gamma_0^2 G'' \tag{8}$$

i.e., the loss modulus is proportional to the energy dissipated per cycle. Using Eqs. (4) and (5), ΔE can be written as

$$\Delta E = \pi \cdot \sigma_0 \cdot \gamma_0 \cdot \sin\delta \approx \pi \cdot \sigma_0 \cdot \gamma_0 \cdot \tan\delta \tag{9}$$

or

$$\Delta E = \pi \cdot \sigma_0^2 \cdot \frac{G''}{|G^*|^2} = \pi \cdot \sigma_0^2 \cdot J'' \tag{10}$$

where J'' is the loss compliance which is defined as $G''/|G^*|^2$. Therefore, depending on whether γ_0, σ_0, or $\gamma_0 \cdot \sigma_0$ is kept constant during dynamic deformation (corresponding to constant strain, constant stress, or constant energy input), the energy loss is proportional to G'', J'', or $\tan\delta$, respectively.

We note that, in principle, the main physical discussions related to filler networking in this paper do not change if a sinusoidal tensile or uniaxial compressional stress (amplitude ε_0) is imposed on the rubber material. In some examples the complex dynamic modulus is then denoted with $E^* = E' + iE''$ and the compliance with $C^* = C' - iC''$. All theoretical considerations use the shearing modulus G^*.

When a sinusoidal strain is imposed on a linear viscoelastic material, e.g., unfilled rubbers, a sinusoidal stress response will result and the dynamic mechanical properties depend only upon temperature and frequency, independent of the type of deformation (constant strain, constant stress, or constant energy). However, the situation changes in the case of filled rubbers. In the following, we mainly discuss carbon black filled rubbers because carbon black is the most widespread filler in rubber products, as for example, automotive tires and vibration mounts. The presence of carbon black filler introduces, in addition, a dependence of the dynamic mechanical properties upon dynamic strain amplitude. This is the reason why carbon black filled rubbers are considered as non-linear viscoelastic materials. The term 'non-linear' viscoelasticity will be discussed later in more detail.

The effect of amplitude-dependence of the dynamic viscoelastic properties of carbon black filled rubbers has been known for some 50 years, but was brought into clear focus by the work of Payne in the 1960s [1–7]. Therefore, this effect is often referred as the Payne-effect. It has been also investigated intensively by

Medalia [8, 9] and Voet and Cook [10, 11]. The effect is illustrated schematically in Fig. 1 for the compression mode. For a specific frequency and specific temperature the storage modulus decreases from a 'zero-amplitude' plateau value, E'_0 (or G'_0 in shearing) to a high-amplitude plateau value, E'_∞ (or G'_∞), with increasing amplitude, whereas the loss modulus shows a pronounced peak.

Other researcher studied the Payne-effect of carbon black filled bromobutyl vulcanizates [12–14], carbon black filled rubbers containing tin end-modified polymers [15], and silica filled rubbers [16–18]. In many cases unusual and smart experimental procedures were applied (e.g., dynamic stress-softening studies [16]). The importance of surface characteristics in the non-linear dynamic properties of filled vulcanizates and the temperature-dependence of the Payne-effect are discussed in a review of Wang [19]. If the deformation amplitude is kept constant, increasing temperature levels lead to decreasing moduli. For theoretical interpretations and modeling it is important to note that the Payne-effect has also been observed in vulcanizates containing polymeric fillers, like microgels of different defined particle size, cross-linking density, and surface chemistry [20, 21].

Even dynamic measurements have been made on mixtures of carbon black with decane and liquid paraffin [22], carbon black suspensions in ethylene vinylacetate copolymers [23], or on clay/water systems [24, 25]. The corresponding results show that the storage modulus decreases with dynamic amplitude in a manner similar to that of conventional rubber (e.g., NR/carbon blacks). This demonstrates the existence and properties of physical carbon black structures in the absence of rubber. Further, these results indicate that structure effects of the filler determine the Payne-effect primarily. The elastomer seems to act merely as a dispersing medium that influences the magnitude of agglomeration and distribution of filler, but does not have visible influence on the overall characteristics of three-dimensional filler networks or filler clusters, respectively. The elastomer matrix allows the filler structure to reform after breakdown with increasing strain amplitude.

It is necessary to state more precisely and to clarify the use of the term 'non-linear' dynamical behavior of filled rubbers. This property should not be confused with the fact that rubbers are highly non-linear elastic materials under static conditions as seen in the typical stress-strain curves. The use of linear viscoelastic parameters, G' and G'', to describe the behavior of dynamic amplitude dependent rubbers may be considered paradoxical in itself, because storage and loss modulus are defined only in terms of "linear" behavior.

When the testing of 'non-linear' rubbers is performed in load contact, harmonics can be expected to show up in the displacement signal due to the non-linear behavior of the materials. However, in many investigations where shearing does not exceed ~100 %, the dynamic mechanical testing devices produce nearly perfectly sinusoidal reference signals, i.e., there are few or no harmonics [26]. Of course, the appearance of intensive harmonics can be observed at larger shear rates realized in liquid-like rubbery melts [27]. If large harmonics were found at small amplitudes, storage and loss modulus would be poor descriptors of the material behavior. Sometimes these discussed features are called 'harmonic paradox' [26]. However, at a given amplitude, the material responds 'lin-

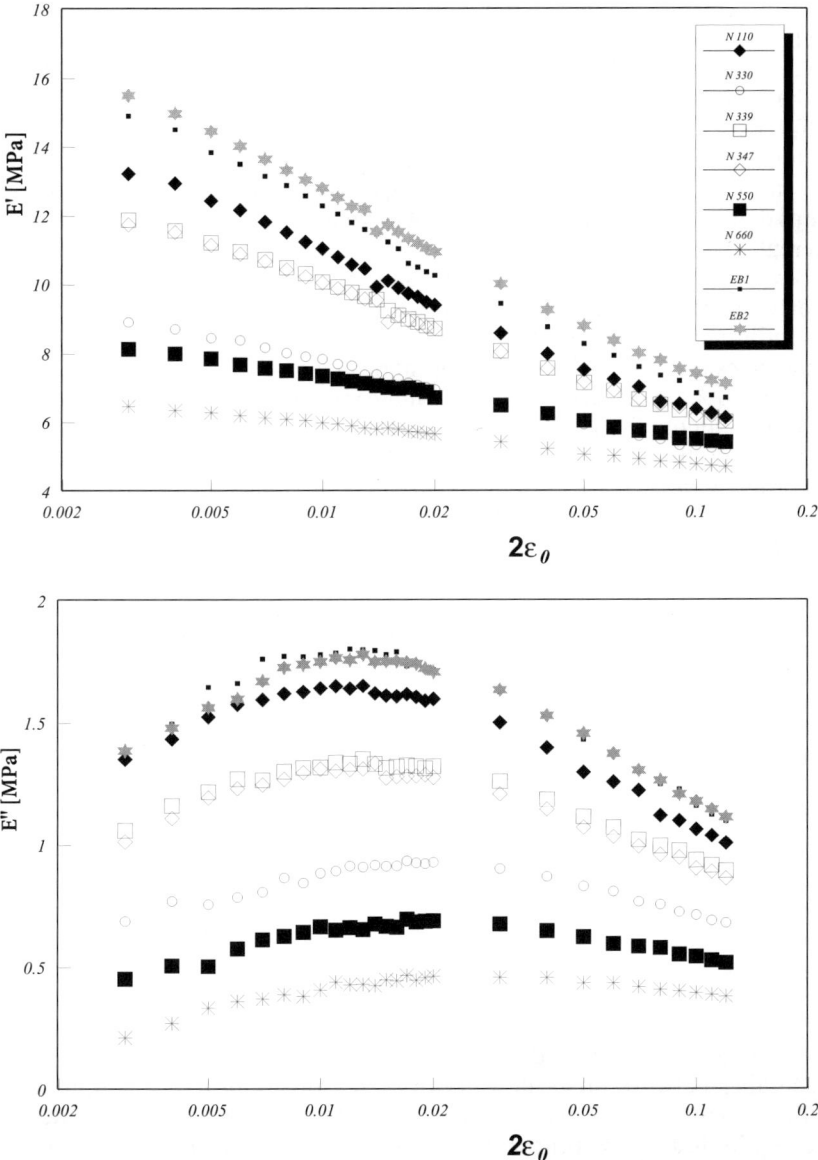

Fig. 1a,b. Strain amplitude dependence of the complex dynamic modulus $E^* = E' + i E''$ in the uniaxial compression mode for natural rubber samples filled with 50 phr carbon black of different grades: **a** storage modulus E'; **b** loss modulus E''. The N numbers denote various commercial blacks, EB denotes non-commercial experimental blacks. The different blacks vary in specific surface and structure. The strain sweeps were performed with a dynamical testing device EPLEXOR at temperature $T = 25\ °C$, frequency $f = 1$ Hz, and static pre-deformation of $-10\ \%$. The x-axis is the double strain amplitude $2\varepsilon_0$

early' and gives a response that is sinusoidal with essentially no harmonics. At a different amplitude, the material again responds 'linearly' but gives an entirely different modulus. Therefore, the storage and loss modulus accurately describe the dynamic mechanical behavior of filled elastomers, despite the dynamic amplitude-dependence.

Another important point is the question whether static offsets have an influence on strain amplitude sweeps. Shearing data show that this seems not to be the case as detailed studied in [26] where shear rates do not exceed ~100 %. However, different tests with low dynamic amplitudes and for different carbon black filled rubbers show pronounced effects of tensile or compressive pre-strain [14, 28, 29]. Unfortunately, no analysis of the presence of harmonics has been performed. The tests indicate that the storage (low dynamic amplitude) modulus E' of all filled vulcanizates decreases with increasing static deformation up to a certain value of stretch ratio λ, say λ^*, above which E' increases rapidly with further increase of λ. The amount of filler in the sample has a marked effect on the rate of initial decrease and on the steady increase in E' at higher strain. The initial decrease in E' with progressive increase in static strain can be attributed to the disruption of the filler network, whereas the steady increase in E' at higher extensions ($\lambda^* \sim 1.2 \ldots 2.0$ depending on temperature, frequency, dynamic strain amplitude) has been explained from the limited extensibility of the elastomer chain [30].

Otherwise, the independence of static offsets at moderate shearing rates can be interpreted in such a way that under such conditions no or only slight disruption of the filler network appears. So far this difference between shearing and tension/compression remains more or less unclear and more experiments are needed to clarify this point. We speculate that under static shearing only slight disruption of the filler network appears and the main effect is only a shear deformation of the shape of filler clusters and its backbone (see Fig. 2).

Payne consequently interpreted the sigmoidal decline from a limiting 'zero-amplitude' value of the storage modulus, G'_0, to a high-amplitude plateau G'_∞ as the result of breakage of physical (van der Waals) bonds between filler particles, or more precisely, the break-down of a secondary aggregate network formed by fillers. A detailed description of fractal structures of the filler network and the structural change with increasing dynamic deformation will be the content of Sects 5 and 6. Payne noted that the value of G'_0 is largely recoverable on return to smaller amplitudes and showed that the phenomenon does not depend on the polymeric nature of the medium [5].

The filler network break-down with increasing deformation amplitude and the decrease of moduli level with increasing temperature at constant deformation amplitude are sometimes referred to as a thixotropic change of the material. In order to represent the thixotropic effects in a continuum mechanical formulation of the material behavior the viscosities are assumed to depend on temperature and the deformation history [31]. The history-dependence is implied by an internal variable which is a measure for the deformation amplitude and has a relaxation property as realized in the constitutive theory of Lion [31]. More qualitatively, this relaxation property is sometimes termed 'viscous coupling' [26] which means that the filler structure is viscously coupled to the elastomeric matrix, instead of being elastically coupled. This phenomenological picture has been successfully introduced in [26] to explain some experimental features like the different frequency dependencies of the limiting moduli G'_0 and G'_∞, etc. The storage modulus at small amplitudes, G'_0, appears to be a function of frequency whereas the storage modulus at high amplitudes, G'_∞, appears to be nearly independent of frequency in a certain range of frequencies (10^{-1} ... 10 Hz) [26]. The maximum of the loss modulus, G''_m, does appear to be a weak function of frequency for most samples with different filler levels [26]. For natural rubber/carbon blacks, the loss modulus at high and low dynamic amplitudes appears to be independent of frequency in the investigated frequency range. Clearly, this point needs further experimental investigations. So far, the phenomenological theory of non-linear thermoviscoelasticity of Lion [31, 32] makes predictions regarding the frequency-dependence of the material parameters which simulate a strain-sweep experiment. The frequency-dependence is fairly weak and in good approximation of a power law type [31, 33].

A very convincing piece of evidence for agglomeration-deagglomeration as an important loss mechanism is the close empirical relationship between the maximum value of the loss modulus, G''_m, and the height of the 'step' in the storage modulus $G'(\gamma_0)$ as a function of deformation amplitude:

$$G''_m = a + b \cdot (G'_0 - G'_\infty) \tag{11}$$

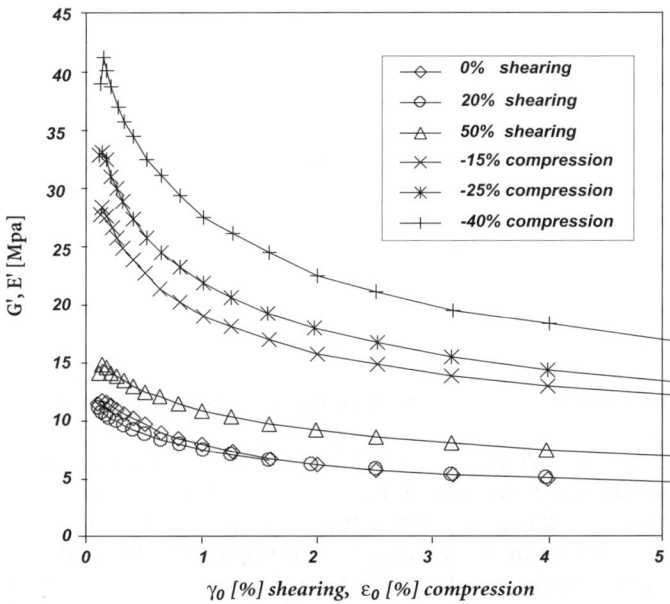

Fig. 2. Comparison of strain sweeps in shearing mode and uniaxial compression mode for different static offsets. Obviously, the influence of offsets is weaker in shearing. The strain sweeps were performed with a dynamical testing device EPLEXOR at temperature T = 25 °C, frequency f = 10 Hz

found by Payne [5] and others [26, 34–36] to hold for a large number of vulcanizates containing both different polymers and carbon blacks in varying concentrations. The empirical constants a and b in Eq. (11) vary somewhat from one rubber to another. The Payne-plot, Eq. (11), does not indicate specifically where a particular rubber material will lie. To determine the location of a specific rubber sample on the plot, a filler effectiveness curve has been introduced, additionally, which shows the change of the storage modulus, $\Delta G' \equiv G'_0 - G'_\infty$, with respect to filler level. Filler effectiveness plot and Payne-plot give a general picture of the relationship between the storage and loss moduli, and strongly suggest that the mechanism for the change in storage modulus and the loss peak are related. However, they do not give information about the nature of the dynamic amplitude-dependence.

Equation (11) will be easily derived in Sect. 2 where we discuss the phenomenological model of Kraus to describe the agglomeration-deagglomeration mechanism. As already noted by Kraus [36], the van der Waals interaction of filler aggregates provides a mechanism of energy storage; however, it does not furnish an obvious loss mechanism. The latter is considered to be provided by excess frictional forces between filler particles or between filler and the polymeric medium as contacts are broken, i.e., each act of breaking and remaking of a contact is deemed to dissipate additional strain energy over and above that dissipated by the vulcanizates in the absence of contacts. These mechanisms can be more brightened within the rubber-junction picture [37] that will be discussed in Sect 3.

In the low strain region (strain < 0.1 %), the dynamic mechanical properties of filled rubbers are less strain-dependent. Payne suggested the presence of a yield point at these low strains [1]. However, Sircar and Lamond indicated that the value of the storage modulus reaches a maximum at an amplitude of around ~0.05 % [35]. Medalia [33] and Payne and Whittaker [38] stated that these results might be attributed to experimental error at very low strain. However, Roland showed that both storage and loss modulus approached constant values in the very low strain region observed with a precise instrumentation [39]. As the strain amplitude rises higher than the order of approximately 0.1 %, strong strain-dependence behavior occurs. The dynamic mechanical properties tend toward 'non-linear' viscoelasticity attributed to the break-down of a filler (agglomerate) network. Beyond approximately >15 % amplitude the filler network is essentially broken down and the dynamic mechanical properties are almost strain-independent [40].

So far we have described and discussed qualitatively the phenomena related to the Payne-effect. We introduced the suggestion that the strain-dependent behavior of the complex modulus G^* is due to the break-down of filler agglomerates (including the infinite cluster, i.e., a throughgoing filler network) as the principal mechanism by which the filler contribute to the energy dissipation in these materials. However, so far no physical model has been introduced here that allows a quantitative description. In the following section we briefly discuss some different approaches to explain the Payne-effect. Moreover, we introduce the seminal phenomenological model suggested by Kraus [36] and some further developments. Critical discussions of advanced models based on fractal characterization of agglomerate and filler network structures will be performed in Sects 5 and 6.

2
The Kraus Model and Related Approaches

The discussion in the Introduction led to the convincing assumption that the strain-dependent behavior of filled rubbers is due to the break-down of filler networks within the rubber matrix. This conviction will be enhanced in the following sections. However, in contrast to this mechanism, sometimes alternative models have been proposed. Gui et al. theorized that the strain amplitude effect was due to deformation, flow and alignment of the rubber molecules attached to the filler particle [41]. Another concept has been developed by Smith [42]. He has indicated that a shell of hard rubber (bound rubber) of definite thickness surrounds the filler and the 'non-linearity' in dynamic mechanical behavior is related to the desorption and reabsorption of the hard absorbed shell around the carbon black. In a similar way, recently Maier and Göritz suggested a Langmuir-type polymer chain adsorption on the filler surface to explain the Payne-effect [43].

Beside physical models a number of authors have proposed pure empirical mathematical models (Dean et al. [44], Martin and Malguarnera [45], Ahmadi and Muhr [46]) or so-called 'frictional' models to describe aspects of the dynamic mechanical behavior of filled elastomers (Resh [47], Fujita et al. [48]). Iwan [49] proposed models consisting of series and parallel combinations of springs to describe yielding behavior in materials. Turner proposed a triboelastic assemblage: a series combination of alternating identical springs and sliders moving relative to a single fixed base [50]. Coveney et al. [51–53] introduced a standard triboelastic solid (STS) model which is the triboelastic equivalent of the standard linear solid [5]. The STS model has been used with some success in finite element analysis and in dynamic characterization of elastomers where emphasis is not generally on (sequential) sinusoidal deformation histories. In practice, elastomer-based components are often subjected to non-sinusoidal histories; for example, 'dual-sine' deformation histories consisting of a primary (higher frequency ω_1, smaller strain amplitude γ_{01}) and simultaneous secondary (lower frequency ω_2, larger strain amplitude γ_{02}) sinusoidal component [55].

Kraus suggested the first phenomenological quantitative model based on agglomeration/deagglomeration of carbon black agglomerates to describe Payne's effect [36]. He assumes that the (van der Waals) carbon black contacts break and reform according to functions – say f_b for breaking and f_r for reforming – of the strain amplitude γ_0. The rate of contact breakage, i.e., the amount of network broken per cycle, R_b, is proportional to the number of existing carbon black contacts and to f_b:

$$R_b = k_b \cdot N \cdot f_b \tag{12}$$

where k_b is the rate constant. The corresponding network reformation rate, R_r, is assumed to be proportional to $N_0 - N$ where N_0 is the number of carbon/carbon contacts at zero deformation:

$$R_r = k_r \cdot (N_0 - N) \cdot f_r \tag{13}$$

k_r being the reformation rate constant. Kraus assumed power laws for the functions f_b and f_r:

$$f_b = \gamma_0^m, \; f_r = \gamma_0^{-m} \tag{14}$$

with m being a constant. Later, we will attribute to m a physical meaning. It will be related to specific fractal dimensions of the fractal agglomerate structures. At equilibrium the two rates are equal ($R_b = R_r$), which gives N as

$$N = \frac{N_0}{1 + \left(\gamma_0 / \gamma_c\right)^{2m}} \tag{15a}$$

where γ_c is a characteristic strain given by

$$\gamma_c = \left(\frac{k_r}{k_b}\right)^{\frac{1}{2m}} \tag{15b}$$

The excess modulus of the agglomeration network at any amplitude over that at infinite strain, $G'(\gamma_0) - G'_\infty$, is taken as proportional to the existing number of contacts N, so

$$\frac{G'(\gamma_0) - G'_\infty}{\Delta G'} = \frac{1}{1 + \left(\dfrac{\gamma_0}{\gamma_c}\right)^{2m}} \tag{16}$$

where $\Delta G' \equiv G'_0 - G'_\infty$. According to the (Cole-Cole-like) function (Eq. 16), γ_c is the amplitude at which $G'(\gamma_0) - G'_\infty$ has decreased to half of its zero-strain value.

As already discussed, Kraus attributes a loss mechanism to be due to excess forces between carbon black particles or between particles and the polymeric medium as contacts are broken [36]. The excess loss modulus may then be taken as proportional to the rate of network breakdown, and

$$G''(\gamma_0) - G''_\infty = c \cdot k_b \cdot \gamma_0^m \cdot N \tag{17}$$

where c is a constant and G''_∞ is $G''(\gamma_0)$ at infinite strain. Substituting for N from Eq. (14) leads with $\Delta G' \sim N_0$ to

$$G''(\gamma_0) - G''_\infty = \frac{C \cdot \gamma_0^m \cdot \Delta G'}{1 + \left(\dfrac{\gamma_0}{\gamma_c}\right)^{2m}} \tag{18}$$

with C being a new constant. The function $G''(\gamma_0)$ has a maximum G''_m at $\gamma_0 = \gamma_c$, so

$$\frac{G''(\gamma_0)-G''_\infty}{G''_m - G''_\infty} = \frac{2\left(\dfrac{\gamma_0}{\gamma_c}\right)^m}{1+\left(\dfrac{\gamma_0}{\gamma_c}\right)^{2m}} \qquad (19)$$

The loss tangent becomes

$$\tan\delta(\gamma_0) = \frac{G''_\infty \left[\left(\dfrac{\gamma_0}{\gamma_c}\right)^{m/2} - \left(\dfrac{\gamma_0}{\gamma_c}\right)^{-m/2}\right] + 2G''_m}{G'_\infty \left(\dfrac{\gamma_0}{\gamma_c}\right)^m + G'_0 \left(\dfrac{\gamma_0}{\gamma_c}\right)^{-m}} \qquad (20)$$

Kraus applied the above equations for storage and loss modulus to data of Payne, finding fairly good correlation. In addition, it was found that m and γ_c are largely independent of polymer, carbon black, and dispersion. This is in accord with the experimentally proved applicability of Eq. (11) which can be easily derived from Eq. (18) when recalling that G''_m occurs when $\gamma_0 = \gamma_c$. Heinrich and Vilgis [56] and Vieweg et al. [57] estimated m from Eq. (20) for a large number of different carbon blacks dispersed in natural rubber and styrene-butadiene copolymers, respectively. They confirmed the universal value of m ($\cong 0.5 \ldots 0.6$) indicating again that it is mainly a geometrical quantity of the filler network and agglomerates, independent of the specific filler type. However, within the Kraus model the exponent m is a pure empirical parameter. The reasons for the universality remaining unclear in this model.

Very recently, Huber and Vilgis proposed a theoretical approach which establishes a connection between the Payne effect and the structural properties of the filler network [58]. They assumed that at very small deformation amplitude the energetically elastic contribution of the rigid filler network is dominant, whereas at higher strains hydrodynamic effect and the rubber-filler interaction dominate. As the filler network breaks up in smaller and smaller units with growing strain, a mean amplitude dependent cluster size $\xi(\gamma_0)$ can be defined. In a first step, this concept of the reinforcing mechanism is mapped on an elementary viscoelastic Zener-like model in a phenomenological way (Fig. 3). The contribution of the filler networking is divided into an elastic and a viscous part, which are both non-linear with respect to the amplitude γ_0 and represent the stored energy and dissipation, respectively, during dynamical break-down of the filler clusters. In a second step the form of the non-linearity is derived from the structure of the filler clusters on the basis of a non-linear relationship between mean cluster size and deformation amplitude. For that purpose Huber and Vilgis [58, 59] adopted some ideas by Witten et al. [60]. By analogy of the Pincus-blob mod-

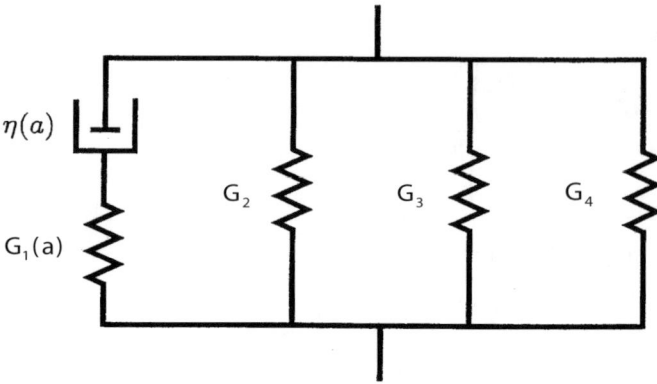

Fig. 3. Visualization of the viscoelastic Zener model within the Huber/Vilgis approach

el from polymer physics [60, 61] Huber and Vilgis identified the blob extension with the (amplitude dependent) mean cluster size, scaling like

$$\frac{\xi}{\xi_0} \sim \gamma_0^{\frac{1}{C-d_f+2}} \qquad (21)$$

if the balance between uniaxial elongation and lateral compression of the filler clusters are taken into account [59]. The agglomeration tendency of fillers like carbon black at higher concentrations leads to the formation of a percolation-like filler network which can be characterized by the fractal dimensions d_f and C, the latter being the connectivity exponent which is related to the minimum path along the cluster structure [62]. In Sect. 6 we will show how in practice the structure of a carbon black network is best described by the model of kinetic cluster-cluster aggregation (CCA) [63, 64]. In this case the fractal exponents take the values $d_f \approx 1.8$ and $C \approx 1.3$. Huber and Vilgis assumed that the viscous part in the generalized Zener model is simply proportional to cluster size, i.e., $\eta(\gamma_0) \sim \xi(\gamma_0) / \xi_0$, whereas the elastic spring turns out to be almost linear, $G_1(\gamma_0) \approx G_1$ [59]. These assumptions improve a previous attempt of Huber et al. to model the Payne effect [65], where the cluster size ξ was supposed to be inversely proportional to the external forces, and the viscosity η was assumed to be constant, i.e., it did not include an explicit mechanism of energy dissipation. As a result, the Huber/Vilgis model [58, 59] yields equations like Eq. (16) for the excess storage modulus and Eq. (19) for the excess loss modulus where

$$G'_\infty = \sum_{i=2}^{4} G_i, \Delta G' = G_1$$

(see Fig. 3). The quantity γ_c in the Kraus model is now a constant collecting remaining system parameters and constants. The exponent m takes the form $m = 1/(C-d_f+2)$ that is entirely determined by the structure of the cluster network. Inserting the special values of d_f and C for CCA clusters leads to $m \approx 0.66$ which is in good agreement with the experimental value $m \approx 0.6$ [56, 57]. The results are not restricted to the case of carbon black fillers (see, for example, [21]); similar behavior for all types of filler particles is expected, independent of their special surface interactions, as long as they form clusters. This feature will be discussed in some more detail below.

We note that the Kraus model provides a fairly good description of the experimental features of the Payne effect. However, very recently Ulmer again evaluated the Kraus equations with data from several published sources and unpublished own data [66]. He found that the description of $G''(\gamma_0)$ according to Eq. (19) is not as good as the description of $G'(\gamma_0)$ according to Eq. (16). The basic deficiency of the Kraus-$G''(\gamma_0)$ model is its inability to account for the G''-values at strains less than about $\sim 10^{-3}$. However, the $G''(\gamma_0)$ description is improved considerably by the addition of a second, empirical term, for example an exponential term like

$$\Delta G''_2 \cdot e^{-\frac{\gamma_0}{\gamma_2}} \tag{22}$$

with $\Delta G''_2$ and γ_2 being two additional unknown parameters. This additional term does not influence the physical ideas leading up to the Kraus model, suggesting that the amplitude-dependence of the dynamic modulus is caused by a thixotropic change in the structure of the filler clusters and the throughgoing filler network. This basic mechanism is also in full line with more recent investigations by Wang et al. [16, 67], who show that the dynamic moduli do not depend only on the current value of the amplitude γ_0, but in the case of strain sweeps, also on the history of deformation. It was found that almost all dynamic stress softening is achieved in the first strain sweep and only minor effects take place in the subsequent sweeps. For all dynamic parameters, the effect of the maximum strain is more important than the number of sweeps [67]. Upon aging at room temperature the softening effect can be partially recovered. Furthermore, it was also found that dynamic stress softening shows some non-universal features like a strong dependence on temperature and frequency, as well as on rubber compounding and filler systems in particular.

In Sect. 1 we discussed the harmonic paradox [26]. At any given amplitude, the material responds 'linearly'. Linear viscoelastic theory predicts, for example, that the storage modulus is low at low frequencies and rises to a maximum value at high frequencies in the dynamical glassy zone. Otherwise, the storage modulus in the dynamic amplitude domains begins high at low amplitudes and decreases as the amplitude increases, but the slope of the respective curves is similar. Linear means that the response of two arbitrary perturbation programs are linear superpositions of the individual responses (Boltzmann's superposition principle). Additionally, the response is causal. This means that the current response (the observable at actual time t) is only influenced by the program in the past, $t' \leq t$. (Note that the Kramers and Kronig dispersion relations being mathematical implications of the linear and causal material equations [54].) Within the linear-response theory it is simple to transform the frequency domain into the time domain through Fourier transforms, and vice versa. However, it is not clear how one would arrive at an amplitude domain from similar arguments. Gerspacher et al. pointed out that the strain-dependency of the modulus G^* can be expressed using a Cole-Cole-like representation $G''=f(G')$ [68]. Then, an equivalence can be made between a mechanical system and an electrical circuit. The '$G''(\gamma_0)$ vs $G'(\gamma_0)$'-plot is similar, at least in shape, to the Cole-Cole type plots obtained by plotting the components ε'' vs ε' of complex dielectric impedance ε^* over a wide range of frequencies [54]. Figure 4 gives an example of Cole-Cole type plots characterizing the Payne effect of different rubbers [56]. Moreover, the mechanical and electrical analogy becomes much more equivalent when replacing G' through G^* and γ_0 through $i\gamma_0$ in the Kraus-equation (Eq. 16) [56]. Then, separating the real and imaginary parts leads to expressions for the storage and loss modulus functions which are related to each other as Hilbert transforms, i.e., through the Kramers-Kronig relations [54]:

$$g^*(\gamma_0) = g'(\gamma_0) + ig''(\gamma_0) \equiv \frac{G^*(\gamma_0) - G'_\infty}{\Delta G'} = \frac{1}{1 + \left(i\dfrac{\gamma_0}{\gamma_c}\right)^\beta} \tag{23}$$

with

$$g'(\gamma_0) \equiv \frac{G'(\gamma_0) - G'_\infty}{\Delta G'} = r^{-1} \cdot \cos\phi \tag{24}$$

$$g''(\gamma_0) \equiv \frac{G''(\gamma_0)}{\Delta G'} = r^{-1} \cdot \sin\phi \tag{25}$$

where

$$r^2 = 1 + 2\left(\frac{\gamma_0}{\gamma_c}\right)^\beta \cdot \cos\left(\frac{\pi}{2}\beta\right) + \left(\frac{\gamma_0}{\gamma_c}\right)^{2\beta} \tag{26}$$

$$\tan\phi = \frac{\left(\dfrac{\gamma_0}{\gamma_c}\right)^\beta \cdot \sin\left(\dfrac{\pi}{2}\beta\right)}{1 + \left(\dfrac{\gamma_0}{\gamma_c}\right)^\beta \cdot \cos\left(\dfrac{\pi}{2}\beta\right)} \tag{27}$$

In the limit of low and high amplitudes, the function $g''(\gamma_0)$ scales, similar to the original Kraus-model, like

$$g''(\gamma_0) \sim \gamma_0^\beta \text{ for } \gamma_0 \ll \gamma_c; \; g''(\gamma_0) \sim \gamma_0^{-\beta} \text{ for } \gamma_0 \gg \gamma_c \tag{28}$$

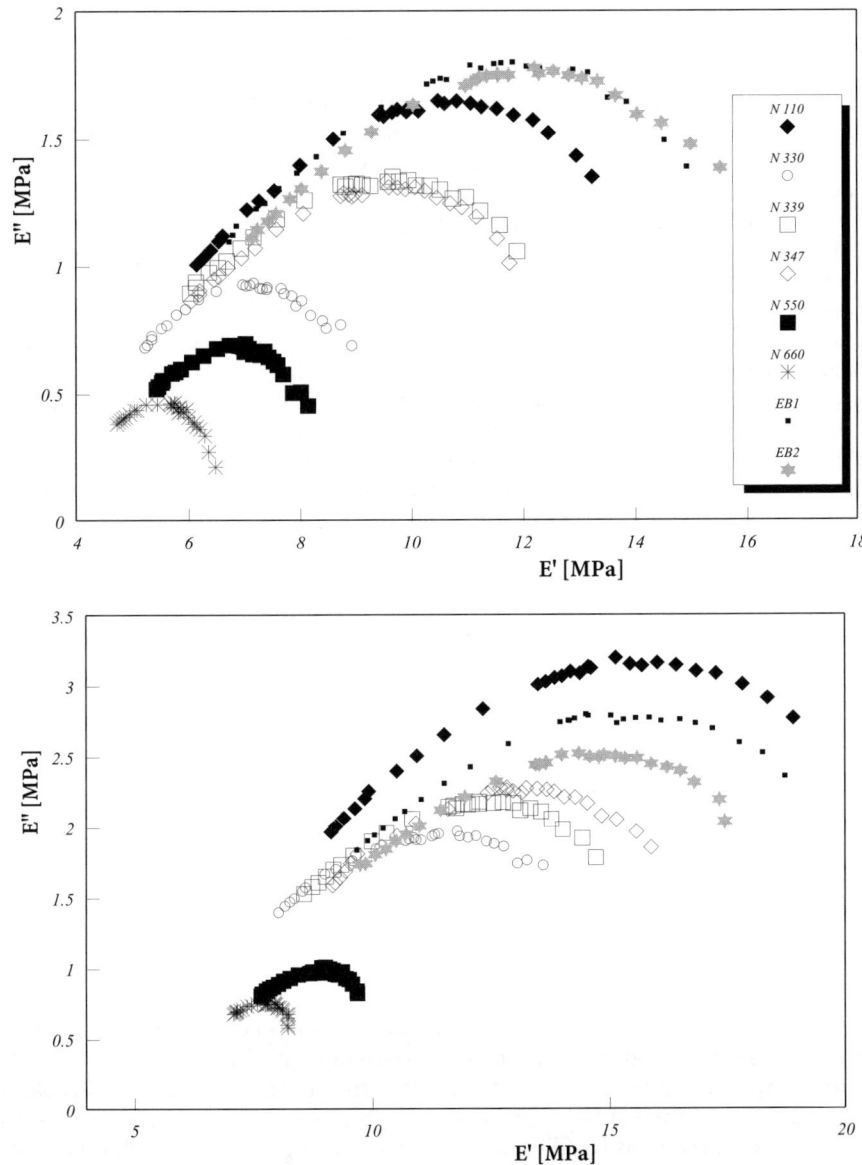

Fig. 4.a Cole-Cole like plots of the strain sweep data from Fig. 1 (polymer matrix: natural rubber). **b** Similar shaped Cole-Cole plots under equal testing conditions in synthetic rubber samples containing the same carbon blacks as indicated in Fig. 4a. The synthetic polymer networks consist of statistical styrene-butadiene copolymers with 23 wt % styrene content (SBR 1500)

Note that Eqs. (23)–(28) exhibit similarities with models for electrical circuits with complex resistors [54]. Physically, this similarity is sometimes related to similarities between momentum transfer in a mechanical system and electron transfer in electrical circuits [69]. In the high strain limit, $\gamma_0 \gg \gamma_c$ where the filler network is nearly totally destroyed, Eqs. (23)–(27) exhibit 'constant-phase-angle' behavior (CPA), i.e., $\phi \cong \beta$ and is independent of the deformation amplitude. In the electrical case the impedance of a CPA element has the form $(i\omega)^{-\beta}$ where ω is the angular frequency, and $0 < \beta < 1$. In recent years it has been demonstrated by many authors that the 'electrical' parameter β is intimately related to surface roughness, with β approaching unity when the surface is made increasingly smooth [70]. Moreover, it has been proposed that general fractional values of β arise from the fractal nature of the rough interface between two materials of very different conductivities, e.g., an electrode and an electrolyte (see, e.g., [70–73] and references cited therein). In a generalized approach, Le Mehauté introduced the so-called TEISI (*Transfer d'Energy sur Interface á Similitude Interne*) model [69] to describe the dynamics of matter and energy transfer on a fractal interface without specifying the nature of transfer at the interface (e.g., response of an arbitrary electrochemical electrode, steady-state transfer across irregular membranes [73], or momentum transfer across fractal interfaces of unconnected carbon black aggregates [68, 74, 75]). The TEISI model, which is a physical expression of the parameterization of the fractal geometry, yields the exponent β as a simple function of the fractal dimension of the self-affine carbon black particle surface: $\beta = 3-D$, where D is the fractal surface dimension. As an example, the parameter β has been estimated for all samples of Fig. 4 via fitting Eq. (27) [56]. The obtained values $\beta \sim 0.4 \ldots 0.6$ yield $D \sim 2.4 \ldots 2.6$. These values of the surface dimension approximately agree with those estimated on pure blacks [76, 77] using neutron scattering and gas adsorption techniques. Furthermore, an universal value of the parameter β confirms the findings that the fractal surface dimensions D of carbon black particles are independent of the carbon black grade [76, 77].

An alternative ability for understanding the CPA-behavior refers to a recently developed micro-mechanical model of stress softening during quasistatic deformations that relates the pronounced hysteresis of filled rubbers to an irreversible breakdown of filler sub-clusters [78, 79]. These sub-clusters are assumed to survive as physically bonded filler units of successively decreasing size up to the rupture strain of the composite. The basic parameter of this model is a strain dependent hydrodynamic amplification factor that describes the hydrodynamic reinforcement of the rubber matrix by the sub-clusters. It is argued that hydrodynamic reinforcement by filler sub-clusters is also a relevant mechanism during dynamical straining at strain amplitudes larger than about 10 % (compare also end of Sect. 6) [78]. This provides a simple explanation of the observed CPA-behavior at large strain amplitudes, since storage and loss modulus are affected by the same hydrodynamic amplification factor that decreases with increasing strain.

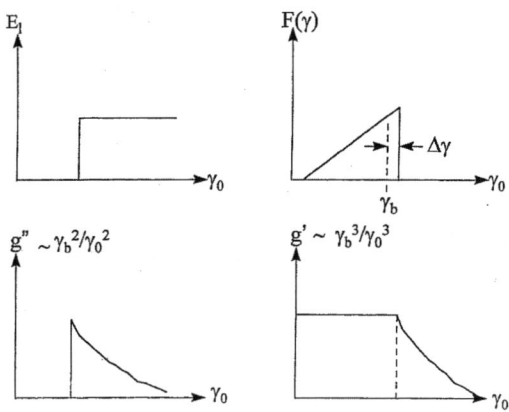

Fig. 5. Visualization of the hysteresis cycle and microscopic complex modulus within the VTG-model. The quantities are explained in the text (from [80])

3
The VTG-Model

Recently, van de Walle, Tricot, and Gerspacher (VTG) modeled the dynamic properties of filled rubbers [80]. In a first step, they assumed a strain-dependence of the storage and loss moduli of a two-aggregate system. Then, the complex moduli of the macroscopic system was derived by introducing a weighting function, $W(\gamma_b)$, γ_b being a critical strain amplitude. It was shown that the relationship between G' and G'' can be established via the weighting function. The strain-dependence of $W(\gamma_b)$ is dependent on the number of interaggregate contacts in the rubber and the strength of interaggregate contacts.

In the VTG-model, two phenomena contribute to the complex moduli of filled rubber compounds: the linear viscoelastic behavior of polymer matrix and the non-linear behavior of the interaggregate contacts. The force between interaggregate contacts is assumed to be of London-van der Waals type. As in the Kraus model it is assumed that, under cyclic deformations, the aggregate contacts are broken and reformed. This leads to an increase in energy loss due to the polymeric chain friction during the separation and reformation of the pair of aggregates. An idealized microscopic model of a pair of aggregates was constructed and its microscopic complex modulus, say $g^* = g' + ig''$, for the idealized model was obtained from the hysteresis cycle as shown in Fig. 5. There, γ_o is the strain amplitude, F is the force between the interaggregate contact, γ_b is the strain amplitude above which the interaggregate contact is broken, and E_1 is the energy loss of a hysteresis cycle.

When $\gamma_o < \gamma_b$, there is no energy loss since the hysteresis does not take place. At the onset of the hysteresis cycle at $\gamma_o = \gamma_b$, the energy loss E_1, i.e., the area enclosed by the hysteresis cycle, can be approximated by

$$E_1 \sim g_0 \cdot \gamma_b \cdot \Delta\gamma \tag{29}$$

where g_o is the slope of the increasing region of force F vs γ_o and $\Delta\gamma$ is the width of the hysteresis cycle. Using the analogy $G'' \sim E_1/\gamma_o^2$, one obtains

$$g''(\gamma_0, \gamma_b) = \begin{cases} 0 & ; if\ \gamma_0 \leq \gamma_b \\ E_1/\gamma_0^2 & ; if\ \gamma_0 > \gamma_b \end{cases} \tag{30}$$

Figure 5 shows g'' as a function of γ_o.

An estimate of the effective g' can be obtained by a linear regression of F vs γ_o on the interval $[0, \gamma_o]$. Assuming that the width of the hysteresis cycle is small, the curve F vs γ_o reduced to the graph $F(\gamma_o)$ as shown in Fig. 5. For $\gamma_o \leq \gamma_b$, the regression yields a slope of g_o, while for $\gamma_o > \gamma_b$ the slope is given by

$$g'(\gamma_o, \gamma_b) \sim g_o \gamma_b^3 / \gamma_o^3 \tag{31}$$

As a result, the microscopic viscoelastic function g^* as a function of γ_o for the idealized model is given as

$$g^* = g_o s'(\gamma_o, \gamma_b) + i g_o \Delta\gamma / \gamma_b s''(\gamma_o, \gamma_b) \tag{32}$$

where

$$\begin{cases} s' = 1 & ; if\ \gamma_0 \leq \gamma_b \\ s' = \gamma_b^3 / \gamma_0^3 & ; if\ \gamma_0 > \gamma_b \end{cases} \tag{33}$$

and

$$\begin{cases} s'' = 0 & ; if\ \gamma_0 \leq \gamma_b \\ s'' = \gamma_b^2 / \gamma_0^2 & ; if\ \gamma_0 > \gamma_b \end{cases} \tag{34}$$

In the composite, there is a distribution of orientation and of separation distances of the interaggregate bonds with respect to direction of the applied strain. As a result, the composite can be considered as a collection of those elementary models, each having a different γ_b. By introducing the weighting function $N(\gamma_b) \cdot d\gamma_b$, which gives the number of links that break when the polymer is stretched from γ_b to $\gamma_b + d\gamma_b$, the complex excess shear modulus of the macroscopic filler system is given by

$$G^*(\gamma_0) = \int_0^\infty g_0(\gamma_b) s'(\gamma_0, \gamma_b) N(\gamma_b) d\gamma_b + i \int_0^\infty g_0(\gamma_b) h s''(\gamma_0, \gamma_b) N(\gamma_b) d\gamma_b \tag{35}$$

where h is the ratio of the width of hysteresis cycle, $\Delta\gamma$, to γ_b and is taken as a constant.

As the effect of a large number of weak links is indistinguishable from that of a smaller number of strong links, g_o and $N(\gamma_b)$ can be combined into a new weighting function $W(\gamma_b)$:

$$W(\gamma_b) = g_o(\gamma_b) N(\gamma_b) \tag{36}$$

Then, if a rubber specific contribution G^*_∞ is added, the complex modulus reduces to

$$G^* = \int_0^\infty s'(\gamma_0, \gamma_b) W(\gamma_b) d\gamma_b + ih \int_0^\infty s''(\gamma_0, \gamma_b) W(\gamma_b) d\gamma_b + G^*_\infty \qquad (37)$$

The real part is

$$G'(\gamma_0) = \int_0^{\gamma_0} \frac{\gamma_b^3}{\gamma_0^3} W(\gamma_b) d\gamma_b + \int_{\gamma_0}^\infty W(\gamma_b) d\gamma_b + G'_\infty \qquad (38)$$

and the imaginary part is

$$G''(\gamma_0) = h \int_0^{\gamma_0} \frac{\gamma_b^2}{\gamma_0^2} W(\gamma_b) d\gamma_b + G''_\infty \qquad (39)$$

The expressions of s' and s'' in Eq. (35) are material independent. On the other hand, $W(\gamma_b)$ is material dependent. In addition, there is a link between G' and G'' because they are derived from the convolution of the same function $W(\gamma_b)$. The link remains across different materials because s' and s'' are material independent.

Equations (38) and (39) indicate that it is easier to estimate $W(\gamma_b)$ from $G''(\gamma_0)$ than from $G'(\gamma_0)$. From Eq. (39) it follows that $W(\gamma_b)$ is related to the differential form of G'':

$$W(\gamma_0) = 1/h[dG''(\gamma_0)/d\gamma_0 + (2/\gamma_0) G''(\gamma_0)] \qquad (40)$$

Equation (40) can now be used to estimate $W(\gamma_0)$ from measurements of $G^*(\gamma_0)$. In a first step one has to subtract the non-filler network contributions, i.e., the high strain storage modulus G'_∞ and the high-strain loss modulus G''_∞, from the values of G' and G'', respectively. Then the following procedures can be used to calculate $W(\gamma_0)$ and optimum values of G'_∞, G''_∞, and h [80]:

$$G''_N(\gamma_0) = G''(\gamma_0) - G''_\infty \qquad (41)$$

$$W(\gamma_0) = 1/h[dG''_N(\gamma_0)/d\gamma_0 + (2/\gamma_0) G''_N(\gamma_0)] \qquad (42)$$

$$G'_N(\gamma_0) = \int_0^\infty s'(\gamma_0, \gamma_b) W(\gamma_b) d\gamma_b \qquad (43)$$

$$G'(\gamma_0)\, cal = G'_N(\gamma_0) + G'_\infty \qquad (44)$$

$$\text{minimize ssd} = (G'(\gamma_0)\, cal/G'(\gamma_0) - 1)^2 \qquad (45)$$

where the subscript N denotes the contribution of filler network. $G'(\gamma_0)$cal is the calculated value of G', and ssd means sum of squares of deviations between experimental and calculated values of G'.

Very recently, Gerspacher et al. [81] applied the VTG-model to study the effect of filler characteristics and loading, filler dispersion, type of filler, type of polymer matrix, and temperature as well as frequency on the dynamic properties by use of the weighting function $W(\gamma_o)$. Figure 6 shows, for example, the plots of the dynamic loss modulus G'' vs γ_o for styrene-butadiene rubber (SBR) vulcanizates filled with different carbon blacks (c. b.) N110, N330, and N660, respectively. The corresponding W-functions according to Eq. (42) are shown in Fig. 7. The weighting functions are used as a measure of the number of interaggregate contacts surviving cyclic deformations. Therefore, as the strain amplitude is increased, a larger number of interaggregate contacts is disrupted and therefore $W(\gamma_o)$ decreases. The calculated values of G' as given in Fig. 8 agree fairly well with the experimental date. Similar results were found for other grades of carbon blacks [81]. In general, the discrepancies between calculated and experimental values of G' are within 1 %, except for strains below 0.2 % where the calculated value of G' is about 6 % higher than experimental values for tread grade carbon blacks, while the discrepancies for carcass grades remain below 1 %.

The VTG-model was formulated with the basic assumption that the reinforcing mechanism of different carbon black grades is similar and only the magnitude of reinforcing differs for different grades. This hypothesis was confirmed by the fact that the Cole-Cole like plots of G'' vs G' for rubber filled with various grades of carbon black were found to be similar through affine transformation [76]. In addition, the Kraus-model parameters m and γ_c, respectively, were found to be ~ 0.6 and ~1.5 %, regardless of grade of carbon black. These universal features have already been discussed in Sect. 2.

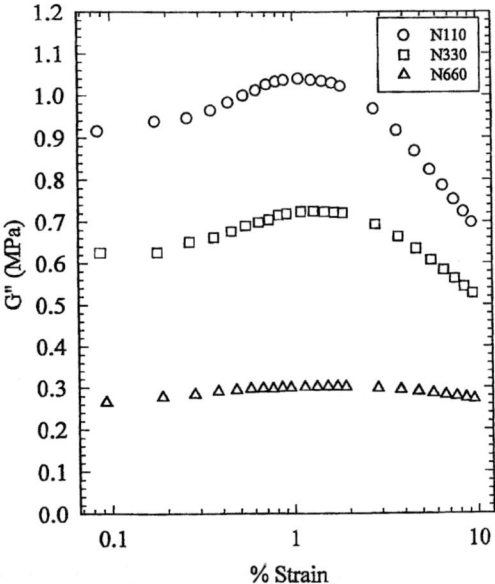

Fig. 6. Strain sweeps for the dynamic loss (shear) modulus G'' of styrene-butadiene rubber samples filled with various carbon blacks (from [81])

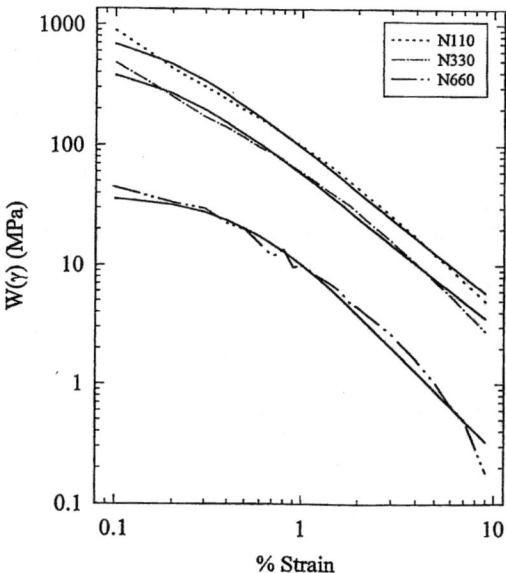

Fig. 7. W(γ)-functions of the VTG-model (Eq. 42) estimated from data points according to Fig. 6 (from [81])

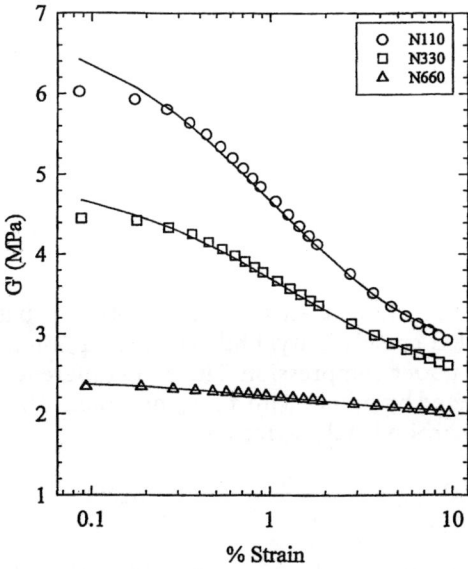

Fig. 8. Calculated G' functions according the VTG-model using the W-functions from Fig. 7 (from [81])

4
The Network Junction (NJ)-Model [37]

Little information has been published on the question of how filler network structure actually affects the energy dissipation process during dynamic strain cycles. The NJ-model focuses on modeling of carbon black network structure and examination of the energy dissipation process in junction points between filler aggregates. This model was further developed to describe the strain amplification phenomenon to provide a filler network interpretation for modulus increase with increasing filler content.

Basic assumptions of the NJ-model are filler-filler attractive forces and the interaggregate distance as the controlling factors for the filler network and dynamic hysteresis that is characterized by either $G''(\gamma_o)$ or $\tan\delta(\gamma_o)$, going through a maximum at medium strain amplitudes. Reduction of $\tan\delta$ can be obtained through improved micro-dispersion of the carbon black, or loosening up the filler network. This may be accomplished, for example, via extensive mixing [83], or by using a coupling agent, such as N,N'- bis (2-methyl-2-nitropropyl)1,6-diaminohexane [84]. Similarly, Wolff et al. [85] showed that in silica filled natural rubber (NR) vulcanizates, the $(\tan\delta)_{max}$-peak can be substantially suppressed by reducing the filler-filler interactions with certain silane coupling agents.

Ouyang et al. started from the total number of contact or junction points for 100 g of carbon black [37]:

$$N_{100} = N_a (\zeta/2) \tag{46}$$

N_a being the total number of aggregates and ζ being the average number of contact points to neighboring aggregates:

$$N_a n_p \left(\frac{\pi}{6}\right) d^3 \varrho = 100 \tag{47}$$

where d is the primary particle diameter, ϱ the carbon black density, and n_p the average number of primary particles in an aggregate. To derive an expression for the junction gap, say h_g, Ouyang et al. made the (critical) assumption that the carbon black network arrangements in filled rubbers is basically the same as those during so-called 'crash' dibutyl phthalate adsorption tests (CDPB) [99] on pure carbon blacks under compression. The main difference is that all junction points are now widened by rubber with a thickness of h_g. The maximum possible volume fraction of carbon black is given by

$$\varphi_{max} = \frac{100/\varrho}{100/\varrho + CDBP} \tag{48}$$

because the minimum amount of rubber needed to replace the void inside a filler network is CDBP. The carbon black volume fraction after all the junction points are expanded by a distance h_g is given by:

$$\varphi = \frac{100/\varrho}{\left\{[100/\varrho+CDBP]^{1/3}+h_g N_{100}^{1/3}\right\}^3} \qquad (49)$$

From Eqs. (46)–(49), one obtains an eypression for the junction gap width as:

$$h_g = \left(\frac{3}{\pi}\right)^{-1/3}\left(\frac{\zeta}{n_p}\right)^{-1/3} P\left[\varphi^{-1/3}-\varphi_{max}^{-1/3}\right]d \qquad (50)$$

with $d \sim 1/(\varrho S)$ and S the specific surface area, e.g., measured by adsorption of cetyltrimethylammonium bromide (CTAB-test) [99]. When carbon black grade and loading are given, that is S, φ, and φ_{max} are fixed, the only remaining factor, that controls the junction widths, is the contact fraction parameter, ζ/n_p. When filler-filler interaction is allowed to be modified, then the contact fraction parameter can have a very powerful effect on h_g (~ relation to surface chemistry of carbon black).

The shape factor for a rubber block, for example a circular disk of diameter d and height h_g, is usually defined as the ratio of one bounded surface to the free surface area. Hence, one obtains with Eq. (50):

$$\frac{\pi d^2}{4\pi d h_g} = \frac{d}{4h_g} = \frac{1}{4}\left(\frac{3}{\pi}\right)^{1/3}\left(\frac{\zeta}{n_p}\right)^{1/3}\left[\varphi^{-1/3}-\varphi_{max}^{-1/3}\right]^{-1} \qquad (51)$$

The shape factor ($\sim d/h_g$) reflects the boundary condition's constraint on rubber flow during deformation, and can be considered as a measure of tightness for a junction. The shape factor, or the ratio d/h_g, can be used to calculate the stored energy with a junction rubber between two spherical filler particles [86, 87]:

$$\Delta\varepsilon' = \frac{1}{2}\Delta f\,\Delta x \sim \frac{1}{2}E_0(\Delta x)^2 d\left(\frac{d}{h_g}\right)^{0.9} \qquad (52)$$

where Δf is the force, Δx the displacement, E_o Young's modulus, d the diameter, and h_g the distance between the two particles.

If it is assumed that for a junction rubber the energy loss takes place near the circular peripherals of the rubber-filler particle interface, the dissipated energy for a cyclic displacement of amplitude Δx and strain rate d/dt (x/h_g) can be estimated to be

$$\Delta\varepsilon'' \sim \eta\frac{1}{h_g}\frac{dx}{dt}\pi d\left(\frac{\Delta x\,d}{4h_g}\right) \qquad (53)$$

where $\eta d/dt$ (x/h_g) represents the stress of the dissipate process with rubber viscosity η. The energy loss could be due to either the actual interfacial slippage between rubber and filler or the rubber internal friction near these regions. This could be the slippage of entanglements (slip-links) in the transition zone between bound rubber layer and mobile rubber phase [88].

When the phase angle in sinusoidal strain cycles is small (i.e., $\sin\delta \approx \tan\delta$), one gets from Eqs. (52) and (53) the loss factor:

$$\tan\delta = \frac{\Delta\varepsilon''}{\Delta\varepsilon'} \sim \left(\frac{\eta}{E_0}\right)\left(\frac{1}{d}\right)\left(\frac{d}{4h_g}\right)^{1.1} \tag{54}$$

The characteristic time $\tau_o = (\eta/E_0)$ represents the effects of modulus as well as hysteresis arising from the rubber and/or the filler-rubber interfacial slippage. The inverse particle diameter, $1/d$, is proportional to the specific surface area, S, and $(d/4\,h_g)^{1.1}$ being the so-called junction shape factor. The proportionality to $h^{-1.1}$ is almost identical to the correlation found by Wolff et al. [89] between $\tan\delta$ and the reciprocal of interaggregate distance, δ_{aa}^{-1}. Combined with Eq. (51), Eq. (54) can be written as

$$\tan\delta \sim \frac{S\left(\dfrac{\eta}{E_0}\right)\left(\dfrac{\zeta}{n_p}\right)^{\frac{1.1}{3}} \varphi^{\frac{1.1}{3}}}{\left[1-\left(\dfrac{\varphi}{\varphi_{max}}\right)^{\frac{1}{3}}\right]^{1.1}} \tag{55}$$

This equation was verified by Ouyang et al. [37] by plotting $\tan\delta$ vs the function $S^* \cdot f(\varphi) \equiv (S/E_o) \cdot \varphi^{(1.1)/3}/[1-(\varphi/\varphi_{max})^{1/3}]^{1.1}$. The result is shown in Fig. 9.

5
The Links-Nodes-Blobs (L-N-B)-Model

Lin and Lee introduced the Links-Nodes-Blob (L-N-B)-model to describe strain-dependent dynamic properties of filled rubber network systems on the basis of a percolation model for the filler network (Fig. 10). The blobs correspond to dense filler aggregates that are not deformed throughout the whole deformation because they are assumed to be entirely rigid. Therefore, a blob can be a primary aggregate or a cluster formed by coagulation of primary aggregates, occluded rubber, and bound rubber. The smallest size of a blob is that of a primary aggregate. The links correspond to tenuous filler clusters between dense filler aggregates and consist of flexible chains of singly connected bonds. The smallest link corresponds to a direct contact bonding between two dense filler aggregates. The links deform under tension, bending, and/or torsion. They even tend to break off when some failure strain is attained. The chains made of links and blobs are called L-N-B chains. The connected points among L-N-B chains are called nodes. The average length of a L-N-B chain between two nearest nodes is defined as ξ_p. Macroscopically, for lengths $>\xi_p$ the system is homogeneous. The length ξ_p corresponds to the critical length

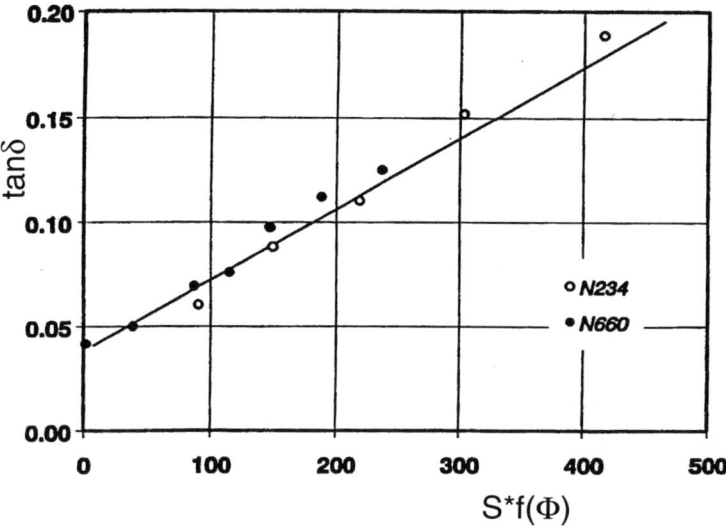

Fig. 9. Linear relationship between tanδ and the function $S^* \cdot f(\varphi)$ (Eq. 55) according the rubber-junction approach. The N numbers denote commercial blacks in the rubber compounds (from [37])

of percolation. Figure 10 shows the equivalent L-N-B model simulating the filler network. The basic assumption of the model is that a blob does not deform and a L-N-B chain breaks only at singly connected bonds on a link.

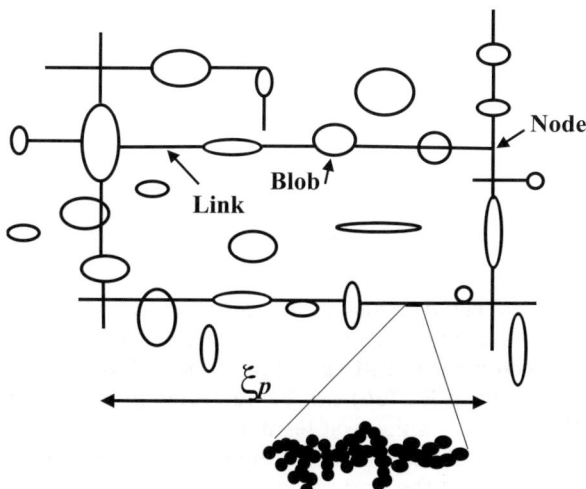

Fig. 10. Visualization of the Links-Nodes-Blob (L-N-B)-model

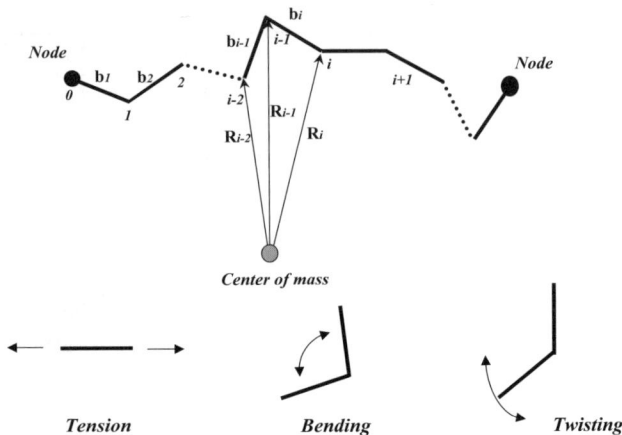

Fig. 11. Filler network chain according the L-N-B-model; i.e., a set of N singly connected bonds under an applied force at the two ends of the chain (after [90])

Lin and Lee started the derivation from Kantor and Webman's two dimensional model of flexible chains that considers a vectorial Born-lattice model with a bending energy term between neighboring bonds [91]. As outlined in Fig. 11, the strain energy H of a chain composed of a set of N singly connected bonds $\{b_i\}$ of length a under an applied force **F** at the two ends of the chain is:

$$H = \frac{F^2 N S_\perp^2}{2G} + \frac{aF^2 L_{ll}}{2Q} \tag{56}$$

where

$$S_\perp^2 = \frac{1}{F^2 N} \sum_{i=1}^{N} \left[(\mathbf{F} \times \mathbf{z})(\mathbf{R}_{i-1} - \mathbf{R}_N) \right]^2 \tag{57}$$

is the squared radius of gyration of the projection of the chain on a two-dimensional plane and

$$L_{ll} = \frac{1}{aF^2} \sum_{i=1}^{N} (\mathbf{F} \cdot \mathbf{b}_i)^2 \tag{58}$$

Here G and Q are local elastic constants corresponding to the changes of angles between singly connected bonds and longitudinal deformation of single bonds, respectively. The vector **z** is a unit vector perpendicular to the plane. For long chains the second term in Eq. (56) can be neglected and the major part of the strain energy H results from the first term, i.e., the bending term of the chain. Then the force constant of the chain relating the elastic energy to the displacement squared of the end of the chain is given by

$$k = G/(NS_\perp^2) \tag{59}$$

In the three-dimensional case ($d = 3$), the angular deformation is not limited to the on-plane bending, but also to the off-plane twisting. To simplify the model the contributions from different kinds of angular deformation are all accumulated in the first term of Eq. (56) by replacing G through an averaged force constant of different kinds of angular deformations $\overline{G}$.

The applicability of Kantor and Webman's model to L-N-B-chains with characteristic, rigid blobs has already been outlined by these authors [91]. It is obtained simply by restricting the summation in Eqs. (57) and (58) over the number L_1 of flexible, singly connected bonds. Then, by introducing the distribution function $f_{1a}(L_1)$ of singly connected bonds and by assuming a strain induced, successive breakdown and a so-called extreme end recombination of bonds during each deformation cycle, the (time averaged) dynamic modulus is evaluated [90]. Without going into further detail, the dynamic moduli of filled rubbers can be written as

$$G'(\gamma_0,\varphi) = G'_F(\gamma_0,\varphi) + G'_\infty(\varphi) \tag{60}$$

$$G''(\gamma_0,\varphi) = G''_F(\gamma_0,\varphi) + G''_\infty(\varphi) + (G''_0(\varphi) - G''_\infty(\varphi)) \int_{\gamma=2\gamma_0}^{\infty} g_{1a}(\gamma)d\gamma \tag{61}$$

where

$$G'_F(\gamma_0,\varphi) = \xi_p^{2-d} \frac{Q\gamma_b}{a^2} \int_{\gamma=2\gamma_0}^{\infty} \frac{f_{1a}(\gamma)}{\gamma} d\gamma, \tag{62}$$

$$G''_F(\gamma_0,\varphi) = \xi_p^{2-d} \frac{Q\gamma_b}{2\pi a^2} \left[\frac{1}{\gamma_0^2} \int_{\gamma=2\gamma_{app}}^{2\gamma_0} \gamma f_{1a}(\gamma) d\gamma \right], \tag{63}$$

$$\gamma_{app} = \left[\frac{mQ + 2\overline{G}}{12\overline{G}} \right] \gamma_b \tag{64}$$

It follows that important physical parameters of the (L-N-B)-model include:
1. The average force constants of a L-N-B chain:

$$k_\xi = \left[\frac{ma^2}{6\overline{G}} + \frac{a^2}{3Q} \right]^{-1} \frac{1}{L_1} \approx \frac{6\overline{G}}{L_1 ma^2} \tag{65}$$

for long chains, where m is the average number of singly connected bonds between two blobs

2. The failure yield strain amplitude of a L-N-B chain:

$$Y_s = \left[\frac{mL_1 Q}{12\overline{G}} + \frac{L_1}{6} \right] \gamma_b \tag{66}$$

where γ_b is the failure strain of a singly connected bond
3. The apparent yield strain amplitude, γ_{app}, that corresponds to $L_1 = 1$ and indicates the onset of non-linearity of the filled rubber
4. The storage and loss modulus at large strain, G'_∞ and G''_∞ that both are controlled primarily by the effective volume of aggregates
5. The critical length of the L-N-B chains, ξ_p
6. The density distribution function of the number of singly connected bonds, $f_{1a}(\gamma)$
7. The density distribution function, $g_{1a}(\gamma)$, that accounts for the break-down of secondary aggregates attributed to non-linear rubber phase deformation.

The local elastic constant $\overline{G}$ is assumed to be controlled by the rubber phase around fillers, i.e., it is primary attributed to bound rubber. The elastic constant Q is controlled by van der Waals forces between fillers. The amplitude gb is the failure strain amplitude for breaking the contact between the constructing particles. Krau [36] derived gb within a soft sphere model as

$$\gamma_b = \frac{\delta_0}{2r}\left(2^{1/3} - 1\right)$$

where do is the equilibrium distance (corresponding to the minimum of the van der Waals potential) between two spheres of radius r. By arbitrarily assuming a primary particle diameter 2r = 30 nm, e.g., for the furnace black N330, and an equilibrium distance do = 0.3 nm one obtains gb = 0.0026. This agrees fairly well with the experimentally observed onset of non-linearity gapp as obtained, e.g., by Payne [1-5], indicating that the bracket term of Eq. (64) is of order unity. The mean number of singly connected bonds between two blobs is expected to be m̄ [a] 2-3 [90]. It implies that the elastic constant Q is almost by one order larger than $\overline{G}$. Note that, dependent on the total number L1 of flexible bonds in a L-N-B chain, the failure strain gs of an L-N-B chain can be significantly larger than the failure strain gb of a single contact due to the bending-twisting ability of the chain (Eq. 66). Since gs varies linearly with L1, one obtains a broad distribution of gs-values that corresponds to the broadness of the distribution function f1a(g) of L1-values of the L-N-B chains. This allows for a natural explanation why the decline of the storage modulus with increasing strain amplitude stretches over more than two decades up to about 100 % strain. The previously discussed models of the Payne effect (Sects 2 and 3), that are based on the rupture of single filler contacts without considering flexible chains or clusters of filler particles, exhibit shortcomings in explaining the broadness of the transition range over more than two decades of strain amplitudes.

It is interesting to note that the (L-N-B)-model leads to similar expressions for the moduli like the VTG-model apart from the first summand of Eq. (38). However, contrary to the semi-empirical weighting functions $W(\gamma_b)$ of the VTG-model, the corresponding density distribution function $f_{1a}(\gamma)$ in the (L-N-B)-model is related to the morphological structure of the filler network, i.e., the distribution of singly connected bonds in a percolation network. Unfortunately, this distribution function is not known, exactly. Therefore, a simple exponential

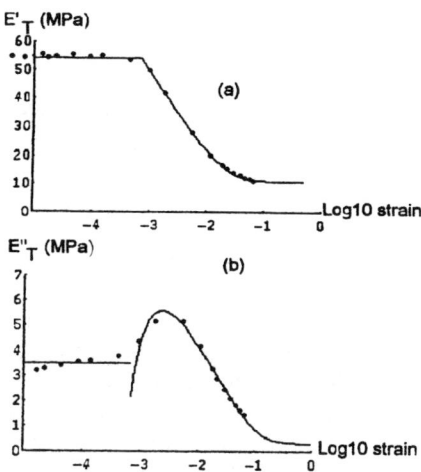

Fig. 12a,b. Fitting functions of storage and loss moduli of carbon black filled natural rubber according the L-N-B-model (from [90])

distribution function with two adjustable parameters is assumed in [90]. The obtained fitting function for the moduli is shown in Fig. 12, where experimental data of carbon black filled natural rubber (NR) are compared to the predictions of the (L-N-B)-model. Typically, the fitted functions are not smooth at the strain amplitude $\gamma = \gamma_{app}$, since the rupture of L-N-B-chains is initiated spontaneously at this particular amplitude followed by an exponential decay of L-N-B-chains with increasing strain. In the large strain regime the fit is fairly good. However, the predicted value of $G''(\gamma_{app})$ is about 50 % smaller than the experimental value.

The (L-N-B)-model represents a highly sophisticated approach to the nonlinear behavior of dynamically excited, filled rubbers. However, it must be criticized that fittings of the storage – and loss – modulus could not be obtained with a single distribution function $f_{1a}(\gamma)$. The physical meaning of the density distribution function $g_{1a}(\gamma)$ remains unclear, indicating that the consideration of energy dissipation in the (L-N-B)-model is uncompleted.

The universal properties of percolation are not considered, properly. In particular, the predicted power law-dependency $G' \sim (p-p_c)^\tau$ with $\tau \approx 3.6$–3.7 is in clear discrepancy with the cited experimental results, where $\tau \approx 1.56$ is obtained [90].

The above described lack of smoothness at $\gamma = \gamma_{app}$ is essential. It refers to the characteristic power law distribution functions of cluster sizes in percolation, indicating that the most frequent number L_1 of singly connected bonds is unity. This leads to a spontaneous fast decline of G' when γ exceeds the value γ_{app}, since all L-N-B-chains with $L_1=1$ break simultaneously at this amplitude. Experimental results show that a smooth transition of G' with varying strain amplitude appears that cannot be described by a power law distribution function or the assumed exponential type of $f_{1a}(\gamma)$.

This leads us to the conclusion that a percolation structure appears inappropriate for the modeling of filler networks in elastomers. Consequently, we will consider an alternative network structure in the next section that refers to a space-filling configuration of kinetically aggregated filler clusters. In particular, this model will be shown to be in agreement with experimental results concerning the effect of filler concentration on the storage modulus.

6
The Cluster-Cluster Aggregation (CCA)-Model

The CCA-model considers the filler network as a result of kinetically cluster-cluster-aggregation, where the size of the fractal network heterogeneity is given by a space-filling condition for the filler clusters [60, 63, 64, 92]. We will summarize the basic assumptions of this approach and extend it by adding additional considerations as well as experimental results. Thereby, we will apply the CCA-model to rubber composites filled with carbon black as well as polymeric filler particles (microgels) of spherical shape and almost mono-disperse size distribution that allow for a better understanding of the mechanisms of rubber reinforcement.

Kinetically cluster-cluster-aggregation (CCA) of filler particles in elastomers is based upon the assumption that the particles are allowed to fluctuate around their mean position in a rubber matrix as detailed studied in [100]. The fluctuation length is comparable to the rubber specific fluctuation length of the chain segments, i.e., the mean spacing of successive chain entanglements. Upon contact of neighboring particles or clusters they stick together, irreversibly, because the thermal energy of colloidal particles is in general much smaller than their interaction energy. Dependent on the concentration of filler particles, this flocculation process leads to spatially separated clusters or a filler network that can be considered as a space-filling configuration of fractal CCA-clusters. The two cases are shown schematically in Fig. 13.

At higher filler concentrations, above the gel point ($\varphi > \varphi^*$), the filler particles come sufficiently close to each other. Under this condition a diffusion

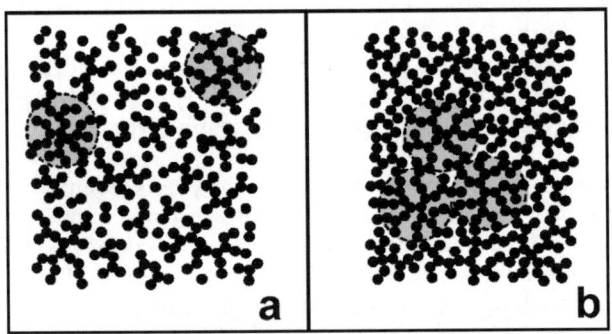

Fig. 13a,b. Schematic view of flocculated filler particles in elastomers: **a** below; **b** above the gel-point φ^* of the filler network. The *shaded circles* indicate individual CCA-clusters

limited cluster by cluster aggregation leads to a space-filling configuration of CCA-clusters, similar to colloid aggregation in low viscosity media [60, 63, 64]. Due to the characteristic self-similar structure of the CCA-clusters with fractal dimension $d_f \approx 1.8$ [93], the cluster growth as described by the solid fraction φ_A of the clusters is given by a space-filling condition, stating that the local solid fraction equals the overall solid concentration:

$$\varphi_A(\varphi) = N_F^{-1}\varphi \text{ for } \varphi > \varphi^* \tag{67}$$

The solid fraction of the fractal CCA-clusters fulfils the scaling law:

$$\varphi_A(\xi) \equiv \frac{N(\xi)d^3}{\xi^3} \cong \left(\frac{d}{\xi}\right)^{3-d_f} \tag{68}$$

Here, N is the number of particles of size d in the clusters of size ξ and N_F in Eq. (67) is a generalized Flory-Number of order one ($N_F \cong 1$) that considers a possible interpenetrating of neighboring clusters [63]. Equations (67) and (68) imply that the cluster size ξ decreases with increasing filler concentration φ according to a power law. This reflects the fact that smaller clusters occupy less empty space than larger clusters (space-filling condition). It means that the size of the fractal heterogeneities of the filler network, i.e., the CCA-clusters shown as dashed circles in Fig. 13(b), decreases with increasing filler concentration.

A necessary condition for rubber reinforcement by filler clusters is the rigidity condition $G_A \gg G_R$, where G_A is the elastic modulus of the clusters and G_R is that of the rubber. This is obvious because a structure that is weaker than the rubber cannot contribute to the stiffening of the polymer matrix. We will see below that the rigidity condition is not fulfilled in all cases because the modulus G_A of the clusters decreases rapidly with increasing size of the clusters. It means that relatively small filler clusters of less than 100 particles can lead to reinforcement of the polymer matrix with $G_R \cong 0.1$ MPa. Here, only this case, necessary for reinforcement, is considered, i.e., the rigidity condition $G_A \gg G_R$ is assumed to be fulfilled. For filler concentrations above the gel point φ^*, where a throughgoing filler network is formed (Fig. 13(b)), stress between the (closely packed) CCA-clusters is transmitted directly between the spanning arms of the clusters that bend substantially. In this case, the strain of the rubber is almost equal to the strain of the spanning arms of the clusters ($\gamma_R \approx \gamma_A$). It means that, due to the rigidity condition $G_A \gg G_R$, the overwhelming part of the elastic energy is stored in the bent arms of the clusters and the contribution of the rubber to the elastic modulus G of the sample can be neglected, i.e., $G \approx G_A$. This indicates that the stored energy density (per unit strain) of highly filled elastomers can be approximated by that of the filler network that in turn equals the stored energy density of a single CCA-cluster. The last conclusion follows from the homogeneity of the filler network on length scales above the cluster size ξ.

For an estimation of the concentration-dependence of the elastic modulus G it is necessary to consider the elastic modulus G_A of the CCA-clusters more closely. By referring to the analytical results, Eqs. (56)–(59) of Kantor and Webman [91], one obtains the elastic modulus of the elastically effective CCA-cluster backbone as the bending-twisting modulus of tender, curved rods [60, 63, 64]:

$$G_A \cong G_P \cdot \left(\frac{d}{\xi}\right)^3 N_B(\xi)^{-1} \cong G_P \cdot \left(\frac{d}{\xi}\right)^{3+d_{f,B}} \cong G_P \cdot (\varphi_A)^{(3+d_{f,B})/(3-d_f)} \qquad (69)$$

Here $G_P = \overline{G}/d^3$ is the averaged elastic bending-twisting modulus of the different kinds of angular deformations of the cluster units, i.e., filler particles or bonds between filler particles. $N_B \cong (\xi/d)^{d_{f,B}}$ is the number of particles in the cluster backbone and $d_{f,B} \approx 1.3$ is the fractal dimension of the CCA-cluster backbone [93]. The last part of Eq. (69) follows from Eq. (68). Equation (69) describes the modulus G_A of the clusters as a local elastic bending-twisting term G_P times a scaling function that involves the size and geometrical structure of the clusters. Consequently, the temperature- or frequency-dependency of G_A is controlled by the front factor G_P or $\overline{G}$. As pointed out in Sect. 5, the local elastic constant $\overline{G}$ is governed by the bound rubber phase around the filler clusters. An essential part of the bound rubber consists of a layer of immobilized, glassy polymer (see below), implying that the temperature- or frequency-dependence of G_A is given by that of the glassy polymer. Since $G \approx G_A$, we expect an Arrhenius temperature behavior for highly filled rubbers that is typically found for polymers in the glassy state. This is in agreement with experimental findings [1]. An example of such Arrhenius-like behavior is shown in Fig. 14. The observed pronounced drop of the storage modulus E' in the temperature range between 20 °C and 70 °C results from the decrease of the stiffness of filler-filler bonds that refers to the bending-twisting energy $\overline{G}$. Obviously, for temperatures above 70 °C the filler clusters become softer than the rubber matrix, i.e., the rigidity condition is no longer fulfilled. Figure 14a indicates that at higher temperatures (>100 °C) the rubber-like elasticity of the polymer network dominates the overall temperature-dependence of the filled rubber sample, i.e., $E' \sim T$. In the temperature range ~ 70 °C to 100 °C both the filler cluster network as well as the polymer network determine the overall temperature behavior. In a somewhat different interpretation the Arrhenius-like behavior is understood by referring it to an apparent strength of single filler-filler bonds constituting the filler network. This strength can be roughly estimated from a filler network cohesion energy, say E_c [103, 104]. The number of sub-units, N (e.g., carbon black aggregates), which are forming a reinforcing network at any temperature, T, can be expressed using the Arrhenius-type relationship. If one uses the Kraus assumption in Sect. 2, that the dynamic modulus G' is proportional to N, one obtains a straight line of slope E_c/R by plotting log G'(T) (or log E'(T)) vs 1/T above the glass-rubber transition region (~ 20 °C to 70 °C) of the rubber sample (Fig. 14), the quantity R being the gas constant.

Extensive studies on different rubber compounds (see, for example, Table 1 in [105]) yield $E_c \sim$ 0.05 to 0.15 eV per filler-filler bond [105, 106], i.e., typical values for physical (van der Waals like) bonds. Similar values were obtained within an approach which assumes a hypothetical analogy between the structure of a statistical carbon black network and that of a Gaussian elastomeric (unfilled) polymer network [107]. As in the Kraus approach, the carbon black network scission process is assumed to be thermally activated.

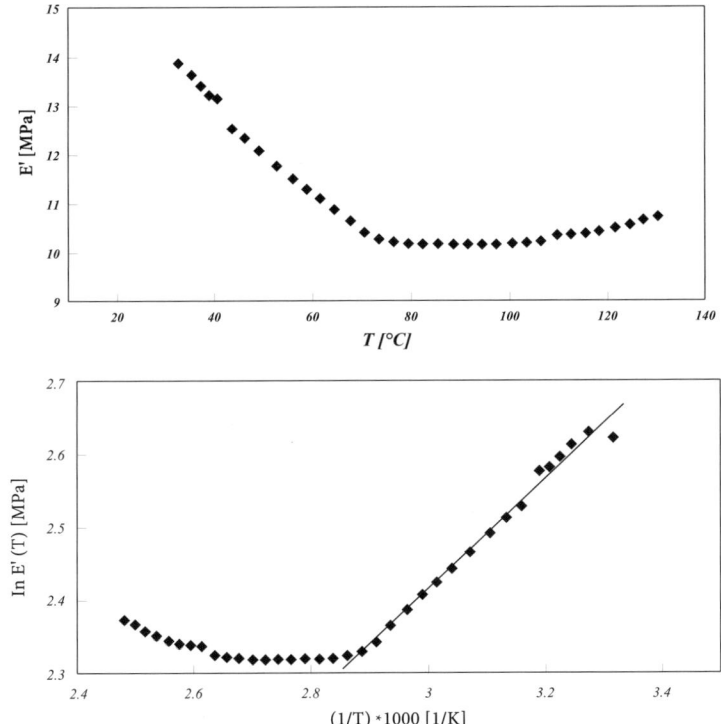

Fig. 14.a Illustration of the Arrhenius-like relationship between storage modulus and temperature above the glass-rubber transition region of a styrene-butadiene rubber (SBR 1500) containing 50 phr carbon black N 339. The temperature sweep was performed with a dynamical testing device EPLEXOR at frequency f = 1 Hz, a static pre-deformation of −10 %, and a dynamical deformation of 1 %. **b** The slope of the linear relation between log E′ and the inverse temperature ($10^3/T \sim 2.9$ to $3.4\ K^{-1}$; corresponding T ~ 20 °C to 80 °C) gives an estimate of the filler networking energy. Data points at larger temperatures indicate that – with increasing temperature – the temperature behavior of the whole filled rubber sample is more and more determined by the entropic elastic behavior of the polymer network (i.e., increasing modulus with increasing temperature)

Table 1: Material parameters from least square fits of Eq. (76) to the experimental data shown in Figure 18 ($\tau = 3.6$)

System	$\Delta G'_0$ [MPa]	γ_c [%]	m [–]	G'_∞ [MPa]
BR(mA)	9.9	44	0.34	1.5
BR(mB)	7.2	34	0.40	1.5
EBR	3.2	117	0.29	0.5
BR	1.0	85	0.31	0.7
N220	8.4	23	0.26	1.1

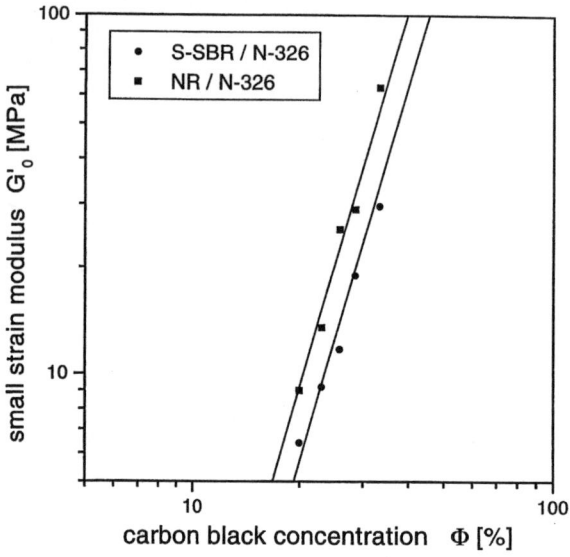

Fig. 15. Double logarithmic plot of the small strain storage modulus vs carbon black volume fraction for a variety of rubber composites as indicated. The *solid lines* with slope 3.5 correspond to the prediction of Eq. (70)

The dependency of the elastic modulus G of the compound on filler volume fraction φ is obtained, if Eq. (69) is combined with Eq. (67):

$$G \cong G_p \varphi^{\frac{3+d_{f,B}}{3-d_f}} \quad \text{for} \quad \varphi > \varphi^* \tag{70}$$

Equation (70) predicts a power law behavior $G \sim \varphi^{3.5}$ for the elastic modulus. Thereby, the exponent $(3 + d_{f,B}) / (3 - d_f) \approx 3.5$ reflects the characteristic structure of the fractal heterogeneity of the filler network, i.e., the CCA-clusters. The predicted power law behavior at higher filler concentrations is confirmed by the experimental results shown in Fig. 15, where the small strain storage modulus of a variety of carbon black filled rubbers is plotted against carbon black loading in a double logarithmic manner. It also agrees with older experimental data obtained by Payne [1] as shown in [63, 64].

Equation (70) is a scaling invariant relation for the concentration-dependency of the elastic modulus of highly filled rubbers, i.e., the relation is independent of filler particle size. The invariant relation results from the special invariant form of the space-filling condition at Eq. (67) together with the scaling invariance of Eqs. (68) and (69), where the particle size d enters as a normalization factor for the cluster size ξ, only. This scaling invariance disappears if the action of the immobilized rubber layer is considered. The effect of a hard, glassy layer of immobilized polymer on the elastic modulus of CCA-clusters can be de-

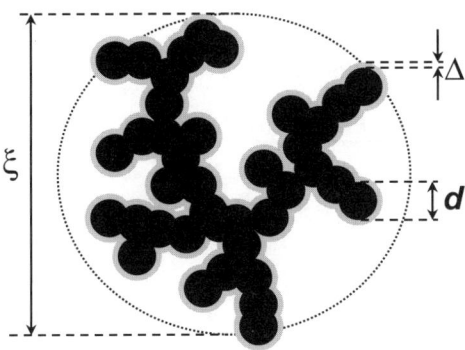

Fig. 16. Schematic representation of the increased solid volume due to an immobilized rubber layer on a filler cluster of spherical colloid particles (ξ: cluster size; d: particle size; Δ: layer thickness)

scribed by introducing a mechanically effective solid fraction of the clusters in analogy to Eq. (68):

$$\tilde{\varphi}_A(\xi) \approx \frac{N(\xi)\left\{\frac{\pi}{6}(d+2\Delta)^3 - \frac{2\pi}{3}\Delta^2\left(3\left(\frac{d}{2}+\Delta\right)-\Delta\right)\right\}}{\frac{\pi}{6}\xi^3} \quad (71)$$

$$\approx \frac{(d+2\Delta)^3 - 6d\Delta^2}{d^3}\varphi_A(\xi) \quad \text{if} \quad \Delta << d$$

This equation considers the mechanically effective solid volume of the clusters, approximately, by enlarging the particle diameter from d to $d+2\Delta$ and subtracting the volume $V = 2\pi/3 \cdot \Delta^2(3(d/2 + \Delta) - \Delta)$ that results from the intersections of the layers of thickness Δ at the contact points between two neighboring particles. The second part of Eq. (71) follows with Eq. (68) and neglects higher order terms in Δ in the second summand. A representation of the increased solid volume due to an immobilized rubber layer is shown schematically in Fig. 16. It becomes obvious that Eq. (71) neglects threefold intersections of the particles that are very close to each other or form small loops. Note, that the definition of effective filler volume fraction according Eq. (71) is qualitatively in accord with definitions suggested by several authors where the effective volume fraction is related to "dead" rubber "trapped" or "caged" in the agglomerates [33, 101]. Even for modeling the reinforcement mechanisms in carbon black and silica loaded (uncrosslinked) rubber melts the effective volume fraction concepts has been introduced successfully [113].

The mechanical action of the immobilized rubber layer on spherical filler particles, that are assumed to form a CCA-filler network in a rubber matrix for $\varphi > \varphi^*$, is obtained if the mechanically effective solid fraction $\tilde{\varphi}_A$ (Eq. (71)), is applied in Eq. (69) instead of φ_A and the space-filling condition $\varphi_A \cong \varphi$ is used.

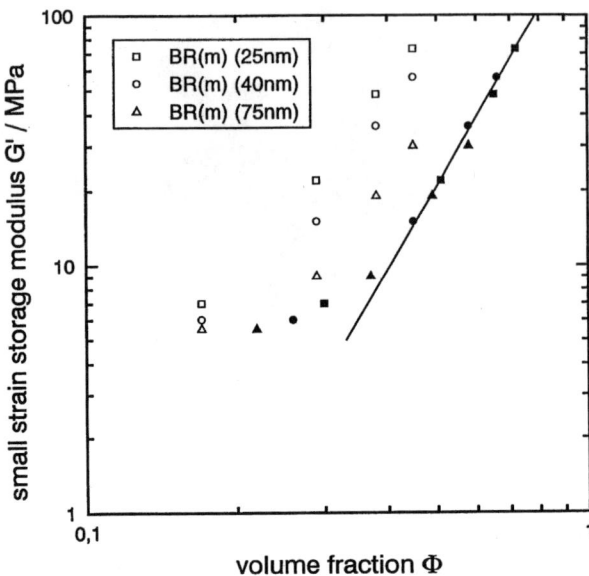

Fig. 17. Double logarithmic plot of the small strain storage modulus vs filler volume fraction for E-SBR/BR(m)-microgel composites with varying size of the BR-microgels as indicated (*open symbols*). The *solid line* represents a master curve with slope 3.5 estimated from Eq. (72) with layer thickness $\Delta = 2$ nm (*solid symbols*). Experimental data are taken from [94, 95]

Then, one obtains the following power law-dependency of the elastic modulus G on filler concentration φ, particle size d, and layer thickness Δ:

$$G \cong G_P \left(\frac{(d+2\Delta)^3 - 6d\Delta^2}{d^3} \varphi \right)^{\frac{3+d_{f,B}}{3-d_f}} \qquad (72)$$

This equation predicts a strong impact of the layer thickness Δ on the elastic modulus G. Furthermore, the influence of particle size d becomes apparent. Obviously, the value of G increases significantly if d becomes smaller, i.e., if the specific surface of the filler increases.

During dynamical excitations of the rubbery material, the quantity G can be identified with the small strain storage modulus G'_o ($G \equiv G'_o$). Figure 17 shows a double logarithmic plot of the small strain storage modulus of variously filled emulsion styrene-butadiene rubbers (E-SBR) vs filler volume fraction (open symbols). Here, the (organic) filler consists of highly cross-linked (with 4 % dicumylperoxide (DCP)), spherical butadiene rubber (BR) microgels of varying size. The size distribution of the filler particles is almost monodisperse. Obviously, G'_o increases with decreasing particle size. This behavior can be related to the increased amount of immobilized rubber with increasing specific surface of the filler particles [94, 95]. All measurement points fall on a single master curve with slope 3.5 (solid line) if an effective volume fraction according to Eq. (72) with a layer thickness $\Delta = 2$ nm, independent of particle size d and concentration φ is considered. The estimated layer thickness $\Delta = 2$ nm that results from a least squares fit appears reasonable. This value agrees with results obtained from NMR measurements [96, 97]. It corresponds to few layers of polymer segments that are fixed at the surface of the microgel clusters like a hard glassy skin. It is important to note that this immobilized rubber layer gives the filler network in polymeric media a higher dynamic stability as compared to colloid networks aggregated in low viscosity liquids. In particular, as shown by Payne [2], a shift of the critical strain amplitude, where the non-linearity related to filler network breakdown occurs, by more than one order of magnitude is observed if the strain amplitude-dependency of the storage modulus of carbon black filled butyl rubber and carbon black filled liquid paraffin are compared. Furthermore, this critical strain amplitude is strongly affected by the surface chemistry of the filler particles that influences the interaction strength between the particle as well as the amount of immobilized rubber.

For a description of the strain-dependency of the storage modulus G' it is convenient to consider $G'(\gamma_0)$ as a sum of two contributions $G'(\gamma_0) = \Delta G'(\gamma_0) + G'_\infty$, where $\Delta G'(\gamma_0)$ describes the strain-dependency of G' that results from a breakdown of the filler network and G'_∞ is the rubber specific contribution including hydrodynamic amplifications due to the presence of the filler. With increasing strain, a random fracture of single clusters in the space filling configuration can be assumed. As broken clusters no longer transmit stress between the remaining clusters of the network, the breakdown of a single cluster is equivalent to removing the cluster from the network configuration. The remaining configuration of CCA-clusters can be handled in the framework of a site percolation network, where occupied sites correspond to CCA-clusters and non-occupied sites correspond to the removed CCA-clusters. Then, the fraction $p(\gamma_0)$ of occupied sites of the percolation lattice equals the ratio between the number $N(\gamma_0)$ of surviving CCA-clusters at given strain γ_0 and the initial number N_o of clusters that is found in the low strain limit $\gamma_0 \to 0$:

$$p(\gamma_0) = \frac{N(\gamma_0)}{N_0} \tag{73}$$

The strain-dependency of the number $N(\gamma_0)$ of surviving clusters can be modeled, in a certain approximation, by the rate at Eq. (15), introduced by Kraus [36]. As already discussed in Sect. 2, the Kraus equation considers a rate equilibrium between the number of broken and reaggregated clusters within each deformation cycle:

$$N(\gamma_0) = \frac{N_0}{1 + \left(\dfrac{\gamma_0}{\gamma_c}\right)^{2m}} \tag{74}$$

Here, γ_c is the strain amplitude, where half of the clusters are broken, m being an empirical exponent. Then, the storage modulus $\Delta G'(\gamma_0) = G'(\gamma_0) - G'_\infty$ at a given strain γ_0 results from Eqs. (73) and (74) and the power law expression for the modulus of a percolation network [62, 64, 91] as follows:

$$\Delta G'(\gamma_0) \cong \Delta G'_0 \cdot \left(p(\gamma_0) - p_c\right)^\tau = \Delta G'_0 \cdot \left[\left(1 + \left(\frac{\gamma_0}{\gamma_c}\right)^{2m}\right)^{-1} - p_c\right]^\tau \quad (75)$$

Here, $p_c \approx 0.2$ is the critical occupation number where a percolation network is formed and $\tau \approx 3.6$ is the elasticity exponent of percolation [91]. For large values of p ($p > p_c$), Eq. (75) can be approximated by a function of the Havriliak-Negami type:

$$\Delta G'(\gamma_0) \cong \Delta G'_0 \cdot \left[1 + \left(\frac{\gamma_0}{\gamma_c}\right)^{2m}\right]^{-\tau} \quad (76)$$

Already with this approximation a fairly good description of the Payne effect is possible. Figure 18 shows experimental results of the strain-dependency of the modulus G' for natural rubber vulcanizates filled with polymeric or conventional carbon black fillers at constant loading of 50 phr [98]. This loading corresponds to a volume fraction $\varphi \approx 0.2$ for the carbon black N220 and to $\varphi \approx 0.33$ for the polymeric fillers. The different polymeric fillers are butadiene rubber microgels (BR(m)) that are highly cross-linked with 2 % DCP. These filler particles have a size of approximately 100 nm. They differ in their surface activity due to epoxidization of 10 % (EBR(m)) and chemical pinning, respectively, that leads to chemically bonded irregular clusters of BR-microgels (BR(mA); BR(mB)) [98]. The solids lines in Fig. 18 correspond to fitted curves according to Eq. (76). The fitting parameters $\Delta G'_0$, γ_c, G'_∞, and m are summarized in Table 1.

It becomes obvious that the small strain modulus $G'_0 = \Delta G'_0 + G'_\infty$ is strongly affected by the structure of the filler particles. Highest values of G'_0 are found for the structured carbon black and the chemically bonded clusters of polybutadiene microgels BR(mA) and BR(mB). A similar behavior is observed for the large strain G'_∞-values that are dominated by the pronounced hydrodynamical amplification of structured particles. Furthermore, the epoxidized EBR(m) shows higher values of G'_0 than the BR(m), which can be related to the stronger interaction between the epoxidized microgel particles. On the other hand, structured filler particles lead to a lower stability of the filler network, as seen by the relatively low γ_c-values for the BR(mA), BR(mB), and N220 samples. The largest γ_c-value is found for the epoxidized EBR(m), suggesting the highest stability corresponds to the case of dipole interactions between the filler particles. The empirical Kraus-parameter m, that considers the power law dependency of the number of surviving clusters on the applied strain, shows a small variation with the applied fillers, only. Its typical value is $m \approx 0.25$–0.4, somewhat lower than the results in [36, 56, 57]. However, the evaluated values for γ_c are significantly higher than those obtained by applying the Kraus-model as considered in Sect. 2.

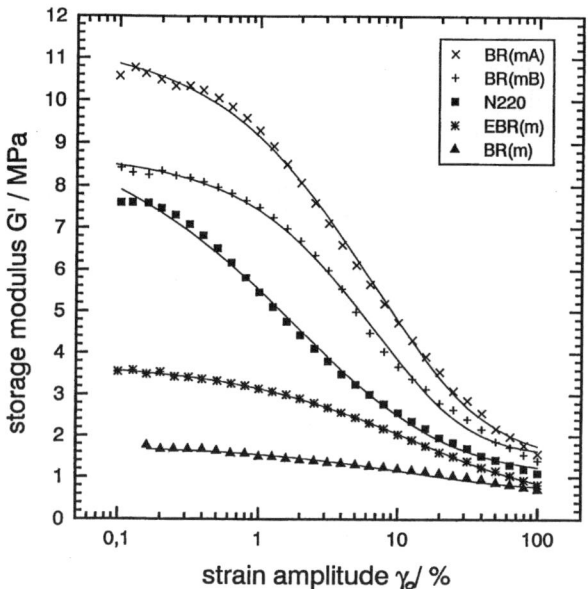

Fig. 18. Plot of the storage modulus vs strain amplitude for NR-vulcanizates filled with 50 phr of a variety of fillers as indicated [98]. The *solid lines* correspond to least square fits according to Eq. (76). Fitting parameters are listed in Table 1

Very recently, the formation of clusters and filler network structures via interparticle interaction has also been observed in composites containing monodisperse size crosslinked polystyrene particles [108].

We finally note that the approach given by Eqs. (73)–(76) suffers from Kraus's somewhat ambiguous statement that the storage modulus $\Delta G'(\gamma_0)$ is a linear function of the number of filler-filler bonds [36], i.e., $\Delta G'(\gamma_0) \sim N(\gamma_0)$ (Eq. 16). This simplified picture is improved by the above physically motivated approach. It is based on the assumption that the initial strong decrease of $G'(\gamma_0)$ with increasing strain γ_0 results from the transition of the filler network to a macroscopically disconnected state with $p(\gamma_0) < p_c$. In order to reach this state, it is not necessary that a large fraction of filler-filler bonds breaks. Instead, the major part of the filler-filler bonds is expected to survive during the initial drop of $G'(\gamma_0)$, observed at relatively small strain amplitudes $\gamma_0 < 100\%$. Most of the remaining sub-clusters break up at significantly larger strains. Parts of the clusters are expected to survive even up to large strains of more than 200 %, causing a subsequent decrease of $G'(\gamma_0)$ even up to the rupture strain. This suggests that an infinite strain amplitude plateau value G'_∞ does not exist in a strict sense.

From this point of view the presented CCA-model is still incomplete. Equations (75) and (76) consider the stored energy density of the filler network that governs the total amount of stored energy as long as $\Delta G'(\gamma_0)$ is significantly larger than the storage modulus G'_R of the rubber matrix. However, according to Eq. (75), $\Delta G'(\gamma_0)$ drops to zero as p approaches p_c, i.e., if $N(\gamma_0) \approx 0.2 N_0$. This means that at large strain amplitudes the contribution of the rubber matrix to the stored energy density becomes dominant. Obviously this contribution is not simply that of the pure rubber. In addition, a hydrodynamic amplification of the rubber matrix by the remaining clusters and broken sub-clusters has to be considered, for example by using the model of Huber and Vilgis. A treatment of such hydrodynamic effects has to include a further breakdown of the sub-clusters with increasing strain. Recently, an exponential decrease of cluster size with increasing strain was considered to obtain a quantitative description of stress softening at large quasistatic strains of various filled rubbers [78, 79] (compare also end of Sect. 2). A complete formulation of strain-induced filler network breakdown during dynamical straining that includes hydrodynamic reinforcement of broken sub-clusters is still outstanding and will be a task of future work. In this context the mechanism of energy dissipation also has to be further clarified.

7
Conclusions

To sum up, for a given polymer system, filler networking plays a critical role in determining dynamic mechanical properties of the elastic filled rubber. Different approaches have worked out that the Payne effect is mainly, if not only, related to the filler network formed in the polymer matrix. Considerable progress has been obtained in the past in relating the pronounced drop of the dynamic elasticity modulus with increasing dynamical strain to a cyclic breakdown and reagglomeration of filler-filler bonds. The breakdown and reformation of the filler network causes an additional energy dissipation hence, higher hysteresis during cyclic strain is expected.

The phenomenological approaches, like the Kraus model or the VTG-model, are found to be well suited for a description of the strain dependence of the complex modulus, though they do not refer to a detailed micro-mechanical model of filler-filler bond dynamics, orientation or strength in a strained system. A more fundamental microscopic basis of non-linear viscoelasticity of reinforced rubbers is provided by the (L-N-B)-model and the CCA-model that consider the arrangement of filler particles in clusters with well-defined fractal structures and the elasticity or fracture of such clusters under external strain. These two models refer to different geometrical arrangements of sub-units in particular filler network structures, resulting from percolation or kinetic cluster-cluster aggregation, respectively.

The established concepts predict some features of the Payne effect, that are independent of the specific types of filler. These features are in good agreement with experimental studies. For example, the Kraus-exponent m of the G' drop with increasing deformation is entirely determined by the structure of the cluster network [58, 59]. Another example is the scaling relation at Eq. (70) predicting a specific power law behavior of the elastic modulus as a function of the filler volume fraction. The exponent reflects the characteristic structure of the fractal heterogeneity of the CCA-cluster network.

The existing concepts of the filler network breakdown and reformation appear to be adequate in describing the deformation-dependence of dynamic mechanical properties of filled rubbers. The different approaches suggest in a common manner that there is a change of filler structure with increasing dynamic strain. This point is worked out fairly well in the existing theories and models. However, in all cases additional assumptions are made about the accompanying energy dissipation process, imparting higher hysteresis to the filled rubber. According to the (NJ)-model the controlling factor of energy dissipation is the filler interaggregate distance. In this picture the breakdown and reformation of the filler network leads obviously to either an excess energy loss due to the interfacial slippage between rubber and filler, or the internal rubber friction near these regions. This may be slippage of entanglements (slip-links) in the transition layer between bound rubber layer and mobile rubber phase [88] (see Sect. 4), and/or partially release of elastically 'dead' immobilized rubber trapped within the filler network or agglomerates [101].

By referring to the (L-N-B)-model, the effect of cyclic breakdown and reaggregation of filler-filler bonds on energy dissipation during each deformation cycle has been evaluated (Sect. 5). It has been found that the estimated energy is – by far – insufficient to explain the macroscopically observed energy dissipation [90]. Nevertheless, the dissipated energy due to the reversibility of cluster breakdown typically goes through a maximum with increasing strain amplitude, which correlates with the maximum of G'', as far as location (strain amplitude) but not magnitude is concerned. The missing energy has been attributed semi-empirically to a breakdown of secondary aggregates via a non-linear rubber phase deformation. A quantitative evaluation of this effect on a microscopic level is outstanding. This confirms our conclusion that at present their is no sufficient micro-mechanical explanation of the markedly enhanced energy dissipation in filled rubber. This remains one of the major tasks for future works.

Useful information about structure and properties of (conducting) carbon black networks in elastomers can be additionally obtained from examinations of the electrical percolation threshold and the dielectric properties in a broad frequency range [63, 64, 102, 109]. In this case, again the fractal structures of the filler network in an 'equivalent circuit model' yield a physical approach to understanding the a.c. behavior of the heterogeneous rubber. In addition, the concept of a local strongly immobilized bound rubber-like layer around the carbon black aggregates (see Sect. 6) is crucial for the explanation of a pronounced dependence of the dielectric permittivity $\varepsilon^*(\omega)$ on the specific surface of carbon blacks [102]. Furthermore, this immobilized bound rubber together with occluded rubber trapped inside the voids of carbon black aggregates explain certain discrepancies between the amount of measured and predicted percolation type power-law exponents for the electrical conductivity of the conductive filler loaded rubbers [110, 111].

We finally note that the concept of filler networking yields a good interpretation of the Payne effect both for filled vulcanized rubber networks and filled uncrosslinked green compounds [112]. The comparison of the G^*-curves of the vulcanizates and of the green compounds for fixed carbon black loading displays an increase of the "G_0-G_∞"-step during vulcanization, i.e., the carbon black aggregates must reagglomerate respectively flocculate during the vulcanization process, which is possible because of the low viscosity of the polymer matrix at high temperatures.

Acknowledgement. G. H. thanks Continental AG for permission to publish the paper.

Note added:
Very recently several electron microscopy studies on elastically stretched nanoparticle chain-like aggregates of inorganic oxides provided direct experimental support for the filler networking concept, i.e. reinforcing mechanism based on an energetics resulting from the breaking of physical bonds holding kinked particle chain segments together (S.K. Friedlander, K. Ogawa, M. Ullmann, J. Pol. Sci.: Part B: Polym. Phys. 38, 2685 (2000); S. K. Friedlander, H. Dong Jang, K. H. Ryu. Appl. Phys. Lett. 72, 173 (1998)).

References

1. Payne AR (1962) J Appl Polym Sci 6:57
2. Payne AR (1963) J Appl Polym Sci 7:873
3. Payne AR (1964) J Appl Polym Sci 8:2661
4. Payne AR (1964) Trans IRI 40:T135
5. Payne AR (1965) In: Kraus G (ed) Reinforcement of elastomers. Interscience Publisher, New York, chap 3
6. Payne AR (1972) J Appl Polym Sci 16:1191
7. Payne AR (1963) Rubber Chem Technol 36:432
8. Medalia AI (1973) Rubber Chem Technol 46:877
9. Medalia AI (1974) Rubber Chem Technol 47:411
10. Voet A, Cook FR (1967) Rubber Chem Technol 40:1364
11. Voet A, Cook FR (1968) Rubber Chem Technol 41:1215
12. Dutta NK, Tripathy DK (1989) Kautsch Gummi Kunstst 42:665
13. Dutta NK, Tripathy DK (1992) J Appl Polym Sci 44:1635
14. Dutta NK, Tripathy DK (1990) Polym Test 9:3
15. Ulmer JD, Hergenrother WL, Lawson DF (1998) Rubber Chem Technol 71:637
16. Wang M-J, Patterson WJ, Ouyang GB (1998) Kautsch Gummi Kunstst 51:106
17. Freund B, Niedermeier W (1998) Kautsch Gummi Kunstst 51:444
18. Mukhopadhyay K, Tripathy DK (1992) J Elastomers Plast 24:203
19. Wang M-J (1998) Rubber Chem Technol 71:520
20. Bischoff A, Klüppel M, Schuster RH (1998) Polym Bull 40:283
21. Vieweg S, Unger R, Heinrich G, Donth E (1999) J Appl Polym Sci 73:495
22. Payne AR, Watson WF (1963) Rubber Chem Technol 36:147
23. Amari T, Mesugi K, Suzuki H (1997) Prog Org Coat 31:171
24. Payne AR, Wittaker RE (1970) Rheol Acta 9:91
25. Payne AR, Wittaker RE (1970) Rheol Acta 9:97
26. Brown JD (1997) Nonlinear dynamic behavior of filled elastomers at small strain amplitudes. PhD Thesis, Rensselaer Polytechnic Institute, Troy, New York; Chazeau L, Brown JD, Yanyo LC, Sternstein SS (2000) Polym Compos 21:202
27. Wilhelm M, Reinheimer P, Orteifer M (1999) Rheol Acta 38:349; (1999) Kautsch Gummi Kunstst 52:754
28. Voet A, Morawski JC (1974) Rubber Chem Technol 47:765
29. Giuliani G, Volpi A (1985) Developments in dynamic testing procedures. Paper No 79, ACS Rubber Division Meeting, Cleveland, Ohio
30. Dutta NK, Tripathy DK, Medalia AI (1973) Rubber World 168:49
31. Lion A (1998) J Mech Phys Solids 46:895
32. Lion A (1999) Rubber Chem Technol 72:410
33. Medalia AI (1978) Rubber Chem Technol 51:437
34. Medalia AI, Laube SG (1978) Rubber Chem Technol 51:89
35. Sircar AK, Lamond TG (1975) Rubber Chem Technol 48:79,89
36. Kraus G (1984) J App Polym Sci, Appl Polym Symp 39:75

37. Ouyang GB, Tokita N, Wang M-J (1995) Paper No 108, ACS Rubber Division Meeting, Cleveland, Ohio
38. Payne AR, Wittaker RE (1971) Rubber Chem Technol 44:440
39. Roland CM, Lee GF (1989) NTIS Rep AD-A2 12824
40. Ulmer JD, Hess WM, Chirico VE (1974) Rubber Chem Technol 47:729
41. Gui KE, Wilkinson CS Jr, Gehmann SD (1952) Ind Eng Chem 44:720
42. Smit PPA (1966) Rheol Acta 5:277
43. Maier P, Göritz D (1998) Kautsch Gummi Kunstst 49:18
44. Dean GD, Duncan JC, Johnson AF (1984) Polym Test 4:225
45. Martin RE, Malguarnera SC (1981) J Elastomers Plast 13:139
46. Ahmadi HR, Muhr AH (1997) Plast Rubber Compos Process Appl 26:451
47. Resh WF (Sept 1990) SAE Tech Paper Ser 901757, Passenger Car Meeting and Exposition, Dearborn, Michigan
48. Fujita T, Suzuki S, Fujita S (1989) ASME, PVP, Seismic Shock Vibration Isolation 181:23
49. Iwan WD (1967) J Appl Mech (ASME) 612
50. Turner DM (1988) Plast Rubber Compos Process Appl 9:197
51. Coveney VA, Johnson DE, Turner DM (1995) Rubber Chem Technol 68:660
52. Coveney VA, Johnson DE (2000) Rubber Chem Technol 73:565
53. Coveney VA, Johnson DE (1999) Rubber Chem Technol 72:673
54. Tschoegl NW (1989) The phenomenological theory of linear viscoelasticity. Springer, Berlin Heidelberg New York
55. Harris JA (1987) Rubber Chem Technol 60:870
56. Heinrich G, Vilgis TA (1995) Macromol Chem Phys Macromol Symp 93:253
57. Vieweg S, Unger R, Schröter K, Donth E, Heinrich G (1995) Polym Network Blends 5:199
58. Huber G, Vilgis TA (1999) Kautsch Gummi Kunstst 52:102
59. Huber G (1997) PhD thesis, University of Mainz, Germany
60. Witten TA, Rubinstein M, Colby RH (1993) J Phys II (France) 3:367
61. De Gennes PG (1979) Scaling concepts in polymer physics. Cornell University Press, Ithaca, London
62. Bunde A, Havlin S (1996) (eds) Fractals and disordered systems. Springer, Berlin Heidelberg New York
63. Klüppel M, Heinrich G (1995) Rubber Chem Technol 68:623
64. Klüppel M, Schuster RH, Heinrich G (1997) Rubber Chem Technol 70:243
65. Huber G, Vilgis TA, Heinrich G (1996) J Phys Condens Matter 8:409
66. Ulmer JD (1996) Rubber Chem Technol 69:15
67. Wang M-J, Patterson WJ, Ouyang GB (1996) Paper No 33, ACS Rubber Division Meeting, Montreal, Canada
68. Gerspacher M, O'Farrell CP (1992) Kautsch Gummi Kunstst 45:97; Gerspacher M (1993) Dynamic viscoelastic properties of loaded elastomers. In: Donnet J-B, Bansal RC, Wang M-J (eds) Carbon black. science and technology. Marcel Dekker, New York Basel Hong Kong
69. Le Méhauté A (1991) Fractal geometries. Theory and applications. CRC Press, Boca Raton Ann Arbor London; Le Méhauté A, Crepy G (1983) Solid State Ionics 9/10:17
70. Sapoval B (1991) Fractal electrodes, fractal membranes, and fractal catalysts. In: Bunde A, Havlin S (eds) Fractals and disordered system. Springer, Berlin Heidelberg New York
71. Liu SH (1985) Phys Rev Lett 55:529
72. Blunt M (1989) J Phys A: Math Gen 22:1179
73. Sapoval B (1994) Phys Rev Lett 73:3314
74. Le Méhauté A, Gerspacher M, Tricot C (1993) Fractal geometry. In: Donnet J-B, Bansal RC, Wang M-J (eds) Carbon black. Science and technology. Marcel Dekker, New York Basel Hong Kong
75. Le Méhauté A (1984) J Stat Phys 36:665

76. Zerda TW, Yang H, Gerspacher M (1992) Rubber Chem Technol 65:130
77. Schröder A, Klüppel M, Schuster RH (1999) Kautsch Gummi Kunstst 52:814; (2000) Kautsch Gummi Kunstst 53:257
78. Klüppel M, Schramm J (2000) Macromol Theory Simul 9:742
79. Klüppel M, Schramm J (1999) An advanced micromechanical model of hyperelasticity and stress softening of reinforced rubbers. In: Dorfmann A, Muhr A (eds) Constitutive models for rubber. A.A. Balkema,, Rotterdam
80. Van de Walle A, Tricot C, Gerspacher M (1994) Paper No 10, ACS Rubber Division Meeting, Pittsburgh, Pennsylvania
81. Gerspacher M, O'Farrell CP, Tricot C, Nikiel L, Yang HA (1996) Paper No 74, ACS Rubber Division Meeting, Louisiana, Kentucky
82. Van de Walle A, Tricot C, Gerspacher M (1996) Kautsch Gummi Kunstst 49:172
83. Welsh FE, Richmond BR, Keach CB, Emerson RJ (1995) Paper No 59, ACS Rubber Division Meeting, Philadelphia
84. Yamaguchi T, Kurimoto I, Ohashi K, Okita T (1989) Kautsch Gummi Kunstst 42:403
85. Wolff S, Wang M-J, Tan E-H (1994) Kautsch Gummi Kunstst 47:102
86. Gent AN, Hwang Y-C (1988) Rubber Chem Technol 61:630
87. Gent AN, Park B (1986) Rubber Chem Technol 59:77
88. Heinrich G, Vilgis TA (1993) Macromolecules 26:1109
89. Wolff S, Wang M-J, Tan E-H (1993) Rubber Chem Technol 61:102
90. Lin C-R, Lee Y-D (1996) Macromol Theory Simul 5:1075; (1997) Macromol Theory Simul 6:339
91. Kantor Y, Webman I (1984) Phys Rev Lett 52:1891
92. Schuster RH, Klüppel M, Schramm J, Heinrich G (1998) Paper No 56, ACS Rubber Division Meeting, Indianapolis
93. Meakin P (1990) Prog Solid State Chem 20:135; (1988) Adv Colloid Interface Sci 28:249
94. Vieweg S (1997) PhD thesis, University of Halle, Germany
95. Vieweg S, Unger R, Hempel E, Donth E (1998) J Non-Cryst Solids 235/237:470
96. Litvinov VM, Steeman PAM (1999) Macromolecules 32:8476
97. Dutta NK, Roy Choudhury N, Haidar B, Vidal A, Donnet J-B (1994) Polymer 35:4293
98. Früh T (1996) PhD thesis, University of Hannover, Germany
99. Hofmann W (1989) Rubber technology handbook. Hanser Publishers, München Wien New York
100. Raos G, Allegra G, Assecondi L, Croci C (2000) Comput Theor Polym Sci 10:149
101. Wang M-J (1999) Rubber Chem Technol 72:430
102. Kastner A, Alig I, Heinrich G, Klüppel M (2002) Polym Bull (in preparation)
103. Fitzgerald ER (1982) Polym Bull 8:331
104. Fitzgerald ER (1982) Rubber Chem Technol 55:1547
105. Heinrich G (1997) Gummi Fasern Kunstst 50:775
106. Heinrich G (1992) Filler-filler interaction. Internal Research Report Continental AG AN92/4.3/20 (unpublished)
107. Susteric Z (1989) Makromol Chem Makromol Symp 23:329
108. Cai JJ, Salovey R (1999) J Mat Sci 34:4719
109. Lanzl T, Ludwig J, Kreitmeier S, Göritz D (2000) Kautsch Gummi Kunstst 53:623
110. Karásek L, Meissner B, Asai S, Sumita M (1996) Polym J 28:121
111. Lin C-R, Chen Y-C, Chang C-Y (2001) Macromol Theory Simul (in press)
112. Niedermeier W (1998) Paper No 28, ACS Rubber Division Meeting, Nashville, Tennessee
113. Eggers H, Schümer P (1996) Rubber Chem Technol 69:253

Received: March 2001

Pharmaceutical Polymeric Controlled Drug Delivery Systems

Majeti N. V. Ravi Kumar[1], Neeraj Kumar[2*], A. J. Domb[2**], Meenakshi Arora[3]

[1] Department of Preventive Medicine and Environmental Health, 354 Health Sciences Research Building, University of Kentucky Medical Center, Lexington, Kentucky 40536, USA
[2] Department of Medicinal Chemistry and Natural Products, The Hebrew University of Jerusalem, P.O. Box 12065, Jerusalem 91120, Israel
[3] Department of Chemistry, Indian Institute of Technology, Roorkee-247 667, India
E-mail: [1]mnvrkumar@yahoo.com, [2*]nk31@mailcity.com, [2**]adomb@cc.huji.ac.il,
[3]meenu_arora@mailcity.com

Drug delivery systems have taken a great impetus to deliver a drug to the diseased lesions. Although this concept is not new great progress has recently been made in the treatment of a variety of diseases. A suitable carrier is needed to deliver a suitable and sufficient amount of the drug to a targeted point, hence, various kinds of formulations are being constantly developed. This paper reviews the present state of art regarding the synthetic methods and characterization of nanoparticles, the suitability of polymeric systems for various drugs, drug loading and drug release properties of various systems such as nanoparticles, hydrogels, microspheres, film and membranes, tablets, etc. The purpose of this review is to summarize the available information so that it will be helpful to beginners and serve as a useful tool for active researchers involved in this area.

Keywords. Controlled drug release, Hydrogels, Microparticles, Nanoparticles, Niosomes, Tablets, Transdermal

List of Abbreviations 47

1 Introduction 49

2 Nanoparticles and Nanospheres 50

2.1   Cross-linking of Amphiphilic Macromolecules 51
2.2   Polymerization of Acrylic Monomers 52
2.3   Polymer Precipitation 53
2.3.1 Solvent Extraction-Evaporation 53
2.3.2 Solvent Displacement or Nanoprecipitation 53
2.3.3 Salting Out 54
2.4   Characterization 54
2.4.1 Particle Size Analysis 54
2.4.2 Surface Charge and Hydrophobicity 54
2.4.3 Methods of Changing Particle Size and Surface Characteristics . . . 55
2.5   Poly(DL-lactide-co-glycolide) Nanoparticles 55
2.6   Poly(ethylene Oxide)-poly(L-lactic acid)/Poly(β-benzyl-L-aspartate) 56
2.7   Poly(lactide-co-glycolide)-[(propylene Oxide)poly(ethylene Oxide)] 56
2.8   Polyphosphazene Derivatives 58

2.9	Poly(ethylene Glycol) Coated Nanospheres	58
2.10	Poly(isobutyl Cyanoacrylate) Nanoparticles	59
2.11	Poly(γ-benzyl-L-glutamate)/Poly(ethylene Oxide) Nanoparticles	59
2.12	Chitosan-polyethylene Oxide Nanoparticles	60
2.13	Methotrexate-O-carboxymethylate Chitosan Nanoparticles	60
2.14	Solid Lipid Nanoparticles	61
3	**Hydrogels and Networks**	**61**
3.1	Cross-Linked Poly(ethylene Glycol) Networks for Protein Delivery	62
3.2	Poly(ε-caprolactone) and Poly(ethylene Glycol) Macromer	62
3.3	Gelatin and Polyacrylamide	63
3.4	Hydroxypropyl Cellulose Gels	63
3.5	Thermally Reversible Xyloglucan Gels	64
3.6	Novel Star-Shaped Gel Polymers	65
3.7	Superporous Hydrogels	65
3.8	Polyelectrolyte Complex Gel of Chitosan and κ-Carrageenan	66
3.9	Chitosan/Polyether Gels	67
3.10	β-Chitin and Poly(ethylene Glycol) Macromer	68
3.11	β-Chitosan/Poly(ethylene Glycol) Macromer	68
3.12	Chitosan-amine Oxide Gel	68
3.13	Poly(ethylene glycol)-co-poly(lactone) Diacrylate Macromers and Chitin	68
3.14	Chitosan/Gelatin Hybrid Polymer Network	69
3.15	Chitosan and D, L-Lactic Acid Hydrogels	70
3.16	Monolithic Gels	70
4	**Niosomes in Drug Delivery**	**70**
4.1	Vesicle in Water in Oil Systems	71
4.2	Niosomes in Hydroxypropyl Methylcellulose	72
4.3	Discomes	72
4.4	Polyhedral Niosomes	74
5	**Microcapsules and Microspheres**	**75**
5.1	Multiporous Beads of Chitosan	75
5.2	Coated Alginate Microspheres	76
5.3	N-(Aminoalkyl)-chitosan	78
5.4	Chitosan/Calcium Alginate Beads	79
5.5	Poly(adipic Anhydride) Microspheres	80
5.6	Gellan-gum Beads	81
5.7	Poly(D, L-lactide-co-glycolide) Microspheres	82
5.8	Alginate-poly-L-lysine Microcapsules	84
5.9	Cross-linked Chitosan Microspheres Coated with Polysaccharides or Lipid	86

5.10	Chitosan/Gelatin Microsphere	87
5.11	Cross-linked Chitosan Network Beads with Spacer Groups	88
5.12	1,5-Dioxepan-2-one (DXO) and D,L-Dilactide (DL-LA) Microspheres	89
5.13	Triglyceride Lipospheres	89
5.14	Glutamate and TRH Microspheres	90
5.15	Polyelectrolyte Complexes of Sodium Alginate/Chitosan	91
5.16	Albumin Microspheres	92
6	**Films and Membranes**	**92**
6.1	Cross-linked Gelatin Films	93
6.2	Transdermal Devices	94
7	**Polymer Tablets**	**95**
7.1	Pectin and Hydroxypropyl Methylcellulose Tablets	95
7.2	Chitosan Tablets	96
7.3	Ethylcellulose Matrix Tablets	96
7.4	Hydroxypropyl Cellulose Tablets	97
8	**Miscellaneous**	**98**
8.1	Mucoadhesive Chitosan Coated Liposomes	98
8.2	Chitosan-Lipid Emulsions	99
8.3	Poly(DL-lactide-*co*-glycolide)-methoxypoly(ethylene Glycol) Copolymers	100
8.4	Polymer Based Gene Delivery Systems	101
8.5	Polypeptides Containing γ-Benzylglutamic Acid	101
8.6	Aliphatic Polyanhydrides	103
8.7	Ricinoleic Acid Based Polymers	105
8.8	Water Soluble Polyamides	106
8.9	Poly(trimethyl Carbonate)-(polyadipic Anhydride) Blends	106
9	**Conclusion**	**107**
References		**108**

List of Abbreviations

AA	Adipic acid
AFM	Atomic force microscopy
BLG	β-Lactoglobulin
BSA	Bovine serum albumin
CDR	Controlled drug release
CF	5(6)-Carboxyfluorescein
CMC	Carboxymethyl cellulose
CNS	Central nervous system

CPM	Chlorphenaramine maleate
DCP	Dicalcium phosphate
DD	Degree of deacetylation
DSC	Differential scanning calorimeter
DPPC	Dipalmitoyl phosphalidylcholine
DXO	1,5-Diozepan-2-one
DL-LA	D,L-Dilactide
IdURD	5-Iodo-2-deoxyuridine
DFS	Diclofenac sodium
DLS	Dynamic light scattering
DVB	Divinylbenzene
EC	Ethylcellulose
5-FU	5-Fluorouracil
FAD	Dimer erucic acid
GPC	Gel permeation chromatography
HPC	Hydroxypropyl cellulose
HEPES	2-[4-[2-Hydroxyethyl]piperazin-1-yl]ethane
^{1}H-NMR	Proton nuclear magnetic resonance
HPMC	Hydroxypropyl methylcellulose
HPN	Hybrid polymer network
IR	Infrared spectroscopy
INH	Isoniazid
mPEG-NH$_2$	Monomethoxy monoamine of PEG
MPS	Mononuclear phagocytic system
MTX	Methotrexate
MCC	Crystalline cellulose
mPEG	Monomethoxypoly(ethylene glycol)
MA	Maleic acid
NFX	Norfloxacin
PEO	Poly(ethylene oxide)
PCS	Photon correlation spectroscopy
PECs	Poly(electrolyte complexes)
PLA	Poly(L-lactic acid)
PBLA	Poly(β-benzyl-L-aspartate)
PLG	Poly(lactide-*co*-glycolide)
PPO	Poly(propylene oxide)
PEG	Poly(ethylene glycol)
PCL	Poly(ε-caprolactone)
PBLG	Poly(γ-benzyl-L-glutamate)
PBS	Phosphate buffered saline
PGA	Poly(glycolic acid)
Poly[MMA]	Poly(methyl methacrylate)
PPG	Poly(propylene glycol)
Poly[HEMA]	Poly(2-hydroxyethyl methacrylate)
PLL	Poly-L-lysine

PAA	Poly(adipic anhydride)
PSA	Poly(sebacic anhydride)
PDA	Poly(dodecanoic acid)
PDMS	Poly(dimethylsiloxane)
PAM	2-Pyridinealdoxime chloride
PTMC	Poly(trimethyl carbonate)
PVA	Poly(vinyl alcohol)
PDMA EMA	Poly[(2-dimethylamino)ethyl methacrylate]
PBDLG	Poly(γ-benzyl-D,L-glutamic acid)
PPG	Poly(propylene glycol)
P(CPP)	1,3-p-Carboxyphenoxypropane
PA	Polyanhydride
PTMG	Poly(tetramethylene glycol)
PACA	Poly(alkyl cyanoacrylate)
QELS	Quasi-elastic light scattering
SEM	Scanning electron microscopy
SIPNs	Semi-interpenetrating polymer networks
SEC	Size exclusion chromatography
SLAB	Sustained local anesthetic blockade
SA	Sebacic acid
Solulan C24	Cholesterol poly-24-oxyethylene ether
Tg	Glass transition temperature
Tm	Melting temperature
TMC	1,3-Dioxan-2-one
TDS	Transdermal drug delivery system
THF	Tetrahydrofuran
TRH	Thyrotropin releasing hormone
TEM	Transmission electron microscopy
Thy-HCl	Thyamine hydrochloride
XRD	X-Ray diffraction
XPS	X-Ray photonelectron spectroscopy

1
Introduction

In recent years, dramatic progress in the area of biomedical engineering has resulted in numerous commercially available macromolecular drugs. Biodegradable polymers have been extensively used in biomedical areas in the form of sutures, wound covering materials, and artificial skin. Polymeric drug delivery systems have been considered for many applications to supplement the standard means of medical therapeutics [1, 2]. These drug delivery systems are less complicated and smaller than the mechanical pumps because the drug can be stored as a dry powder within the polymer matrix. Recent advances have shown that polymeric devices are useful for high molecular weight drugs [2] and for those drugs that should be delivered in minute quantities with zero-order kinetics [3].

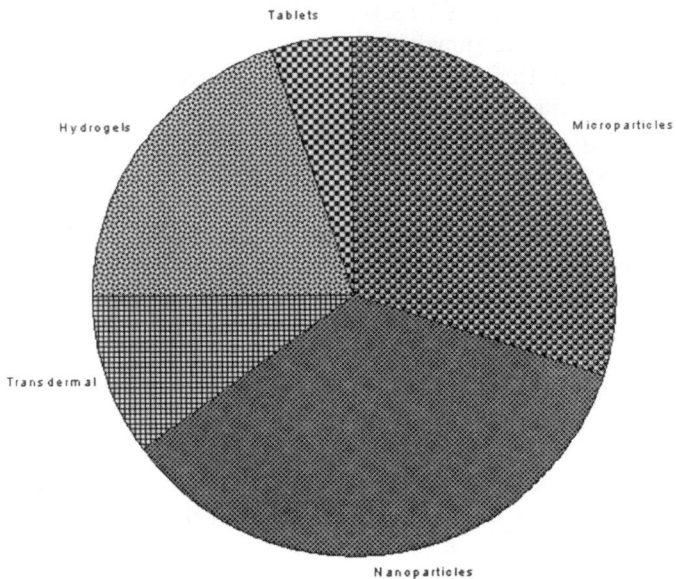

Fig. 1. Controlled drug delivery by various routes

New controlled drug delivery systems that respond to changes in environmental conditions, e.g., temperature [4], pH [5], light (ultraviolet [6] or visible [7]), electric field [8], and certain chemicals [9] are being explored. Such systems, which are potentially useful for pulsed drug delivery, experience changes in either their structures or their intra-intermolecular interactions due to external stimuli as mentioned above [10]. Controlled drug release formulations have been tried in various forms depending upon the end use specification, the major forms being nanoparticles followed by microparticles and hydrogels (Fig. 1). Of the several formulations described herein, nanoparticles have made a tremendous impact in CDR technology, therefore it is worthwhile to describe the preparation and characterization techniques. This article is compiled in such a way that it serves as a useful tool for beginners as well as those who are actively involved in this fascinating area of drug delivery technology.

2
Nanoparticles and Nanospheres

Nanoparticles were first developed around 1970 and are defined as solid colloidal particles, less then 1 μm in size, that consist of macromolecular compounds. They were initially devised as carriers for vaccines and anticancer drugs [11]. The use of nanoparticles for ophthalmic and oral delivery was also investigated [12]. Drugs or other biologically active molecules are dissolved, entrapped, or

Table 1. Methods for the preparation of nanoparticles based on candidate drug and polymers used

Nanoparticulate technique	Polymer type	Candidate drug
i. Emulsion polymerization	Hydrophobic	Hydrophilic
ii. Interfacial O/W polymerization	Poly(alkyl cyanoacrylates)	Hydrophobic (soluble in oils)
iii. Solvent extraction-evaporation	Polyesters (PLA, PLGA, PECL)	Hydrophilic/hydrophobic
iv. Solvent displacement		Soluble in polar solvent
v. Salting-out		Soluble in polar solvent
i. Heat denaturation and cross-linking in W/O emulsion	Hydrophilic	Hydrophilic
ii. Desolvation and cross-linking in aqueous medium	Albumin, gelatin	Hydrophilic and protein affinity
iii. Cross-linking in aqueous medium	Alginate, chitosan, dextran	Hydrophilic and protein affinity
iv. Polymer precipitation in an organic solvent		Hydrophilic, not soluble in polar solvent

encapsulated in the nanoparticles or are chemically attached to the polymers or adsorbed to their surface.

Several methods have been developed for preparing nanoparticles and are optimized on the basis of their physicochemical properties (e.g., size and hydrophilicity) with regard to their in vivo fate after parenteral administration. The selection of the appropriate method for preparing drug-loaded nanoparticles depends on the physicochemical properties of the polymer and the drug. On the other hand, the procedure and the formulation conditions will determine the inner structure of these polymeric colloidal systems. Two types of systems with different inner structures are possible:

(i) a matrix-type system composed of an entanglement of oligomer or polymer units, defined here as a nanoparticle or nanosphere, and
(ii) a reservoir type system, consisting of an oily core surrounded by a polymer wall, defined here as a nanocapsule.

We can classify the preparation methods for the formation of nanospheres on the basis of the material used as shown in the Table 1.

In spite of the reviews published on the preparation methods of nanoparticles [13,14], we feel that a brief description of these methods would be more significant in the present context.

2.1
Cross-linking of Amphiphilic Macromolecules

Nanoparticles have been prepared from polysaccharides, proteins, and amphiphilic macromolecules by inducing their aggregation followed by stabilization either by heat denaturation or chemical cross-linking. The former can be done by water-in-oil emulsion system or in aqueous environments. The cross-linking technique was first used by Cramer et al. in 1974 [15]. In this technique, an aqueous solution of protein was emulsified in oil using a high-speed homogenizer/sonicator and the water-in-oil emulsion was then poured into a hot oil having a temperature greater than 100 °C and held for a specific time (to denature the protein), thereby leading to the formation of submicroscopic particles. These particle were finally washed with organic solvents and subsequently collected by centrifugation. The crucial factors in the technique of nanoparticle production were emulsification energy and stabilization temperature; however, the limitation of the high stabilizing temperature was overcome by adding a chemical cross-linking agent, e.g., glutaraldehyde to the system. To achieve a variable size of the nanospheres, several formulations were adopted and optimized. Protein and polysaccharides nanoparticles can be obtained by a phase separation process in aqueous medium. This can be induced by desolvation of the macromolecule, by a change in pH or temperature, or by adding a counterion in the acid medium [18].

2.2
Polymerization of Acrylic Monomers

PACA nanoparticles have been prepared by an emulsion polymerization method in which droplets of water-insoluble monomers are emulsified in an aqueous and acidic phase that contains a stabilizer. The monomers polymerize relatively rapidly by an anionic mechanism, the rate of polymerization being dependent on the pH of the medium. The system is maintained under magnetic agitation while the polymerization reaction takes place. The duration of polymerization reaction is determined by the length of the alkyl chain, varying from 2 to 12 h for ethyl and hexyl cyanoacrylate, respectively. Finally, the colloidal suspension is neutralized and lyophilized following incorporation of glucose as a cryoprotectant. Water-soluble drug may be associated with PACA nanospheres either by dissolving the drug in the aqueous polymerization medium or by incubating blank nanospheres in an aqueous solution of the drug. The drug loading efficiency is dependent on various vectors, including the pK_a and polarity of the drug, size and surface charge of the nanospheres, and the drug concentration in the aqueous medium [19].

In another method of encapsulation of lipophilic drugs into PACA polymers, the monomers and the drug have been dissolved in a mixture of a polar solvent (acetone or methanol), an oil (coconut oil or benzyl benzoate) and a lipophilic surfactant, such as lecithin. The organic phase is added into an aqueous phase containing a hydrophobic surfactant (e.g., Poloxamer 188) under magnetic agitation. Thus, diffusion of the polar solvent into aqueous phase and the polymerization of the monomer at the oil-water interface take place simultaneously. Polymerization is initiated by the hydroxide ions and leads to the formation of nanocapsules having an oily core surrounded by a polymer coat. The organic solvent is eliminated completely from the colloidal suspension. The selection of the oil plays a major role as it influences the size of the nanocapsules, the molecular weight of the polymer coat, and the stability of the suspension after storage [20].

2.3
Polymer Precipitation

Solvent precipitation techniques have been generally applied for hydrophobic polymers, except for dextran nanospheres. Several techniques described in the literature are based on the mechanism of polymer precipitation.

2.3.1
Solvent Extraction-Evaporation

In this technique, a hydrophobic polymer is dissolved in an organic solvent, such as chloroform, ethyl acetate, or methylene chloride and is emulsified in an aqueous phase containing a stabilizer (e.g., PVA). Just after formation of the nanoemulsion, the solvent diffuses to the external phase until saturation. The solvent molecules that reach the water-air interphase evaporate, which leads to continuous diffusion of the solvent molecules from the inner droplets of the emulsion to the external phase; simultaneously, the precipitation of the polymer leads to the formation of nanospheres. The extraction of solvent from the nanodroplets to the external aqueous medium can be induced by adding an alcohol (e.g. isopropanol), thereby increasing the solubility of the organic solvent in the external phase. A purification step is required to assure the elimination of the surfactant in the preparation. This technique is most suitable for the encapsulation of lipophilic drugs, which can be dissolved in the polymer solution.

2.3.2
Solvent Displacement or Nanoprecipitation

In this method, the organic solvent selected is completely dissolved in the external aqueous phase, thus there is no need of evaporation or extraction for polymer precipitation. Polymer and drug are dissolved in acetone, ethanol, or methanol and incorporated under magnetic stirring into an aqueous solution of the surfactant.

The organic solvent diffuses instantaneously to the external aqueous phase, followed by precipitation of the polymer and drug. After formation of the nanoparticles, the solvent is eliminated and the suspension concentrated under reduced pressure. The advantage of this method is that no surfactant is employed; however, the method is limited to drugs that are highly soluble in a polar solvent.

2.3.3
Salting-Out

This technique, based on the precipitation of a hydrophobic polymer, is useful for the encapsulation of either hydrophilic or hydrophobic drugs because a variety of solvents, including polar (e.g., acetone or methanol) and non-polar (methylene chloride or chloroform) solvents can be chosen for dissolving the drug. This procedure is just like nanoprecipitation, however, the miscibility of both phases is prevented by the saturation of the external aqueous phase with PVA. Precipitation occurs when a sufficient amount of water is added to allow complete diffusion of the acetone into the aqueous phase.

2.4
Characterization

Nanoparticles as colloidal carriers mainly depend on the particle size distribution, surface charge, and hydrophilicity. These physicochemical properties affect not only drug loading and release, but also the interaction of these particulate carriers with biological membranes.

2.4.1
Particle Size Analysis

Two main techniques have been used to determine the particle size distribution of colloidal systems: PCS and electron microscopy including both SEM and TEM. The QELS technique for Brownian moment measurement, offers an accurate procedure for measuring the size distribution of nanoparticles. The PCS technique does not require any particular preparation for analysis and is excellent due to its efficiency and accuracy. However, its dependency on the Brownian movement of particles in a suspended medium may affect the particle size determination.

Electron microscopy provides an image of the particles to be measured. In particular, SEM is used for vacuum dried nanoparticles that are coated with a conductive carbon-gold layer for analysis and TEM is used to determine the size, shape, and inner core structure of the particles. TEM in combination with freeze-fracture procedures differentiates between nanocapsules, nanospheres, and emulsion droplets. AFM is an advanced microscopic technique and its images can be obtained in aqueous medium. AFM images, nowadays are a powerful support for the investigation of nanoparticles in biological media.

2.4.2
Surface Charge and Hydrophobicity

The interaction of nanospheres with a biological environment and electrostatic interaction with biological compounds occur due to the charge on the surface, e.g., a negative charge promotes the adsorption of positively charged drug molecules such as aminoglucosides as well as enzymes and proteins. The surface charge of colloidal particles can be determined by measuring the particle velocity in an electric field. Nowadays, laser light scattering techniques, in particular laser Doppler anemometry, are fast enough to measure the surface charge with high resolution. The hydrophobicity of the nanoparticles can be determined by methods including the adsorption of hydrophobic fluorescent or radiolabeled probes, two phase partitions, hydrophobic interaction chromatography, and contact angle measurements. Recently, XPS has been developed which offers the identification of chemical groups in the 5 Å-thick coat on the external surface of the nanospheres. Gref et al. [21] have characterized the PEG-coated PLGA nanosphere and identified the PEG chemical elements that were concentrated on the nanosphere's surface.

2.4.3
Methods of Changing Particle Size and Surface Characteristics

The fate of colloidal particles inside the body depends on three factors: particle size, particle charge, and surface hydrophobicity. Particles with a very small size (less than 100 nm), low charge, and a hydrophilic surface are not recognized by the mononuclear phagocytic system (MPS) and, therefore, have a long half-life in the blood circulation. In general, nature and concentration of the surfactant play an important role in determining the particle size, as well as the surface charge, e.g., nanospheres with a mean size of less than 50 nm were prepared by increasing the concentration of Poloxamer 188 [22]. The approaches for modifying surface charge and hydrophilicity were initially based on the adsorption of hydrophilic surfactants, such as block copolymers of the poloxamer and poloxamine series. The in vivo studies of hydrophilic nanospheres are of limited usefulness due to their toxicity on intravenous injection. Recently, the idea of using diblock copolymers made of PLA and PEO has been widely accepted owing to safety and stability of the hydrophilic coat. For this purpose, the copolymer is dissolved in an organic solvent, and then emulsified in an external aqueous phase, thereby orienting the PEO toward the aqueous surrounding medium, while in another method the PLA-PEO copolymer is adsorbed on to preformed PLGA nanoparticles. This was found to be efficient in prolonging the nanosphere's circulation time following intravenous administration.

2.5
Poly(D,L-lactide-*co*-glycolide) Nanoparticles

The most widely used emulsion solvent evaporation method for preparation of nanoparticles using PLGA requires surfactants to stabilize the dispersed particle [23]. This method often has a problem that the surfactant remains at the surface of the particles and is then difficult to remove when PVA is used as surfactant. Other surfactants such as the span series or tween series, PEO, etc. are also used for stabilization but have some disadvantages like removal of solvents, toxicity, low particle yield, consumption of more surfactant, and multi-step processes.

A most important factor needing to be considered when using surfactants is that they are non-biodegradable, non-digestible, and moreover, tend to induce allergic reactions in humans. Recently, Jeon et al. [24] proposed a surfactant-free method for the preparation of PLGA nanoparticles as an alternative.

The surfactant-free PLGA nanoparticles were prepared by a dialysis method using various solvents and their physiochemical properties were investigated with regard to the used solvent. Release kinetics of NFX showed that higher drug contents tend to larger particle sizes and slower release [25].

Yoo et al. [26] reported the antitumor activities of nanoparticles based on a doxorubicin-PLGA conjugate via the ester linkage that is expected to be more readily cleavable under physiological conditions. They have studied the antiumor activity in vivo by a subcutaneous route in comparison to the daily injection of doxorubicin and found that doxorubicin-PLGA conjugates are potentially useful for the treatment of tumors [26].

Santos-Magalhaes et al. [27] reported on PLGA nanocapsules/nanoemulsions for benzathine pencillin G. Nanoemulsions were produced by spontaneous emulsification and nanocapsules by interfacial deposition of the pre-formed polymer. They have observed similar release kinetics from both formulations [27].

2.6
Poly(ethylene oxide)-poly(L-lactic acid)/poly(β-benzyl-L-aspartate)

Polymeric micelles are expected to self-assemble when block copolymers are used for their preparation [28]. Micelles of biocompatible copolymer, viz., PEO with PLA or with PBLA, have been reported in the literature [29, 30]. The synthesis of such nanospheres with functional groups on their surface is shown in Fig. 2.

Aldehyde groups on the surface of the PEO-PLA micelles might react with the lysine residues of cell's proteins and for attachment of the amino-containing ligands. These hydroxy groups on the surface of the PEO-PBLA micelles can be further derivatized and conjugated with molecules able to pilot the modified micelles to specific sites of the living organism. Such nanospheres have been tested as vehicles for the delivery of anti-inflammatory and anti-tumor drugs [31, 32].

2.7
Poly(lactide-*co*-glycolide)-[(propylene Oxide)-poly(ethylene Oxide)]

Biocompatible and biodegradable PLG nanoparticles (80–150 nm) have been prepared by following the nanoprecipitation technique [33]. The nanoparticles were coated with a 5–10 nm thick layer of PPO-PEO block copolymer or with tetrafunctional $(PEO-PPO)_2N-CH_2-CH_2-N(PPO-PEO)_2$ [33]. Such coats are bound to the core of the nanosphere by hydrophobic interactions of the PPO chains, while the PEO chains protrude into the surrounding medium and form

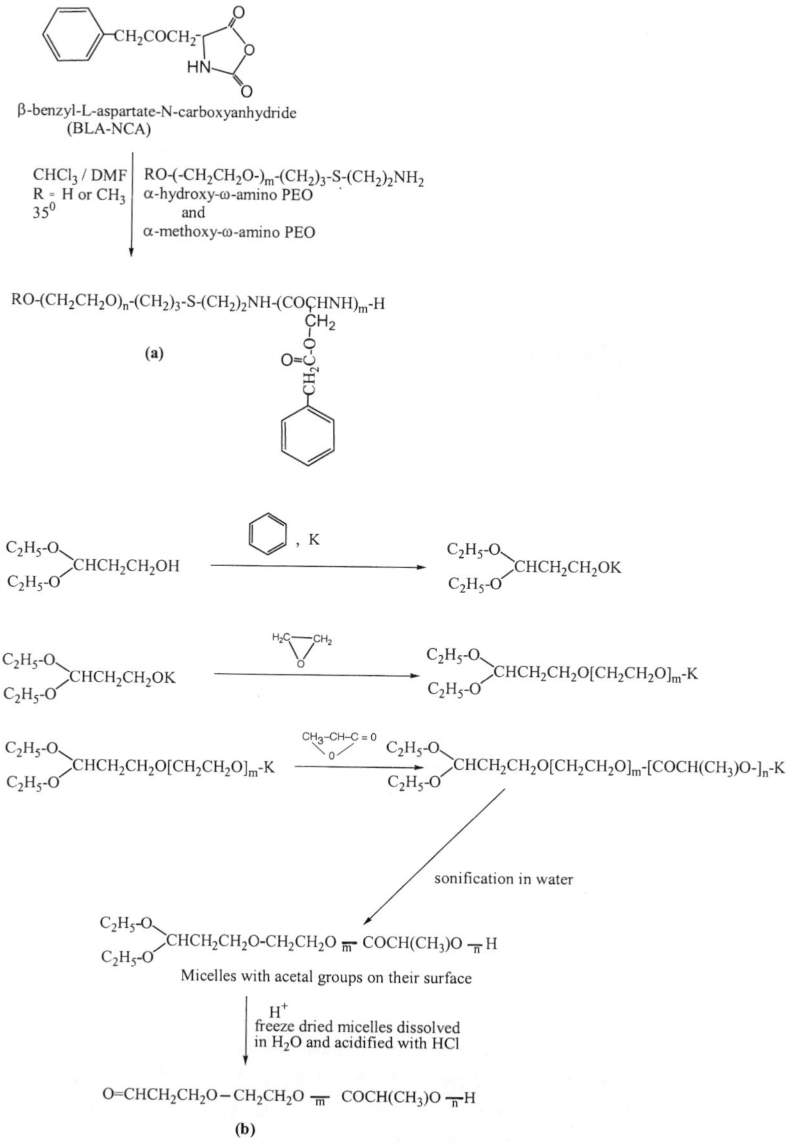

Fig. 2. (*a*) PEO-PBLA and (*b*) PEO-PLLa micelles with aldehyde groups on their surface

a steric barrier, which hinders the adsorption of certain plasma proteins onto the surface of such particles. On the other hand, the PEO coat enhances the adsorption of certain other plasma compounds. In consequence, the PEO-coated nanospheres are not recognized by macrophages as foreign bodies and are not attacked by them [34].

2.8
Polyphosphazene Derivatives

Allcock and coworkers developed derivatives of the phosphazene polymers suitable for biomedical applications [35, 36]. Long-circulating in the blood, 100–120 nm in diameter, PEO-coated nanoparticles of the poly(organophospazenes) containing amino acids have been prepared. The PEO-polyphosphazene copolymer or poloxamine 908 (a tetrafunctional PEO copolymer) has been deposited on their surface [37].

2.9
Poly(ethylene Glycol) Coated Nanospheres

Nanospheres of PLA, PLG, or PCL coated with PEG may be used for intravenous drug delivery. PEG and PEO denote essentially identical polymers. The only difference between the respective structures is that methoxy groups in PEO may replace the terminal hydroxy groups of PEG. PEG coating of nanospheres provides protection against interaction with the blood components, which induce removal of the foreign particles from the blood. PEG coated nanospheres may function as circulation depots of the administered drugs [21, 38]. Slow release of the drugs into plasma alters the concentration profiles leading to therapeutical benefits. PEG coated nanospheres (200 nm), in which PEG is chemically bound to the core have been prepared, in the presence of monomethoxy-PEG, by ring opening polymerization (with stannous octoate as a catalyst) of monomers such as ε-caprolactone, lactide, and/or glycolide [38]. Ring opening polymerization of these monomers in the presence of multifunctional hydroxy acids such as citric or tartaric, to which several molecules of the monomethoxymonoamine of PEG (MPEG-NH$_2$) have been attached, yields multiblock (PEG)$_n$-(X)$_m$ copolymers. It has been demonstrated that morphology, degradation, and drug encapsulation behavior of copolymers containing PEG blocks strongly depend on their chemical composition and structure. Studies of nanoparticles composed of the diblocks of PLG with the methoxy-terminated PEG (PLG-PEG) or of the branched multi-blocks PLA-(PEG)$_3$, in which three methoxy-terminated PEG chains are attached through a citric acid residue, suggested that they have a core corona structure in aqueous medium. The polyester blocks form the solid inner core. The nanoparticles, prepared using equimolar amounts of the PLLA-PEG and PDLA-PEG stereoisomers, are shaped as discs with PEG chains sticking out from their surface.

Their hydrophobic/hydrophilic content seems to be just right for applications in cancer and gene therapies. Such nanospheres are prepared by dispersing the methylene chloride solution of the copolymer in water and allowing the solvent to evaporate [38]. By attaching biotin to the free hydroxy groups and complexation with avidin, cell-specific delivery may be attained. NMR studies of such systems [39] revealed that the flexibility and mobility of the thus attached PEG chains is similar to that of the unattached PEG molecules dissolved in water. Recently, PLG microspheres, with the PEG-dextran conjugates attached to their surface, have been investigated as another variant of the above-described approach. Microspheres with diameters of 400–600 nm have been prepared [40]. Targeting moieties can be attached to the glycopyranose hydroxy groups of the dextran units. Stella et al. [41] proposed a new concept of folic acid conjugated nanoparticles for drug targeting. The authors claimed that these nanoparticles represent potential carriers for tumor cell-selective antitumoral drugs.

2.10
Poly(isobutyl Cyanoacrylate) Nanocapsules

Intragastric administration of insulin-loaded poly(isobutyl cyanoacrylate) nanocapsules induced a reduction of the glycemia to normal level in streptozotocin diabetic rats [42] and in alloxan-induced diabetic dogs [43]. The hypoglycemic effect was characterized by surprising events including a lag time period of 2 days and a prolonged effect over 20 days. Insulin is a very hydrosoluble peptide and should be inactivated by the enzymes of the gastrointestinal tract. Thus, the reason why insulin could be encapsulated with high efficiency in nanocapsules containing an oily core and why these nanocapsules showed such unexpected biological effects remained unexplained. Nanocapsules were prepared by interfacial polymerization of isobutyl cyanoacrylate [44]. Any nucleophilic group including those of some of the amino acids of insulin [45] could initiate the polymerization of such a monomer. In this case, insulin could be found covalently attached to the polymer forming the nanocapsule wall as was recently demonstrated with insulin-loaded nanospheres [46].

Aboubakar et al. [47] studied the physico-chemical characterization of insulin-loaded poly(isobutyl cyanoacrylate) nanocapsules obtained by interfacial polymerization. They claimed that the large amount of ethanol used in the preparation of the nanocapsules initiated the polymerization of isobutyl cyanoacrylate and preserved the peptide from a reaction with monomer, resulting in a high encapsulation rate of insulin. From their investigations, it appears that insulin was located inside the core of the nanocapsules and not simply adsorbed onto their surface.

Lambert et al. [48] used poly(isobutyl cyanoacrylate) nanoparticles for the delivery of oligonucleotides. Nanoparticles of size ranging from 20–400 nm were prepared. The authors claimed that this technology might offer interesting perspectives for DNA and peptide transport and delivery.

2.11
Poly(γ-benzyl-L-glutamate)/Poly(ethylene Oxide)

Hydrophilic-hydrophobic diblock copolymers exhibit amphiphilic behavior and form micelles with a core-shell architecture. These polymeric carriers have been used to solubilize hydrophobic drugs, to increase blood circulation time, to obtain favorable biodistribution, and to lower interactions with the reticuloendothelial system [49]. In the same direction, Oh et al. [50] reported the preparation and characterization of polymeric nanoparticles containing adriamycin as a model drug. The nanoparticles were obtained from the PBLG/PEO diblock copolymer, with the form of a hydrophobic inner core and a hydrophilic outer shell of micellar structure [51, 52] by adopting a dialysis procedure. Their results indicate that only 20 % of the entrapped drug was released in 24 h at 37 °C and the release was dependent on the molecular weight of the hydrophobic polymer.

2.12
Chitosan-poly(ethylene Oxide) Nanoparticles

Hydrophilic nanoparticle carriers have important potential applications for the administration of therapeutic molecules [28, 53]. Most of the recently developed hydrophobic-hydrophilic carriers require the use of organic solvents for their preparation and have a limited protein-loading capacity [54, 55]. Calvo et al. [56] reported a new approach for the preparation of nanoparticles, made solely of hydrophilic polymer, to address these limitations. The preparation technique, based on an ionic gelation process, is extremely mild and involves the mixing of two aqueous phases at room temperature.

One phase contains the polysaccharide chitosan (CS) and a diblock copolymer of ethylene oxide and the polyanion sodium tripolyphosphate (TPP). It was stated that the size (200–1000 nm) and zeta potential (between + 20 mV and + 60 mV) of nanoparticles can be conventionally modulated by varying the ratio of CS/PEO to PPO. Furthermore, using BSA as a model protein, it was shown that these new nanoparticles have a high protein loading capacity (entrapment efficiency up to 80 % of the protein) and provide a continuous release of the entrapped protein for up to 1 week [56].

2.13
Methotrexate-*O*-carboxymethylate Chitosan

Nanoparticles of methotrexate (MTX; 4-amino-4-deoxy-*N*-methylfolic acid) were prepared using *O*-carboxymethylate chitosan (*O*-CMC) as wall forming materials and an isoelectric-critical technique under ambient conditions [57]. The controlled release of drugs was studied in several media including simulated gastric fluid, intestinal fluid, and 1 % fresh mice serum. It was found that acidic media provide a faster release than neutral media. The effects of the MTX/*O*-CMC ratio and the amount of cross-linking agents on drug release in different media were evaluated. The changes of size and effective diameter of the *O*-CMC nanoparticles were detected by SEM and laser light scattering before and after the drug release. The author claimed that the *O*-CMC nanoparticles constitute an attractive alternative to other anticancer drugs and enzyme carriers [57].

2.14
Solid Lipid Nanoparticles

Solid lipid nanoparticles (SLNs), one of the colloidal carrier systems, has many advantages such as good biocompatibility, low toxicity, and stability [58]. Schwarz and Mehnert [59] studied the lipophilic model drugs tetracaine and etomidate. The study highlights the maximum drug loading, entrapment efficacy, and effect of drug incorporation on SLN size, zeta potential (charge), and long-term physical stability. Drug loads of up to 10 % were achieved with a good maintenance of physically stable nanoparticle dispersion [59]. They claimed that the incorporation of drugs showed no or little effect on particle size and zeta potential compared to drug-free SLN [59]. In another study, Kim and Kim [60] examined the effect of drug lipophilicity and surfactant on the drug loading capacity, particle size, and drug release profile. They prepared SLNs by homogenization of melted lipid dispersed in an aqueous surfactant solution. Ketoprofen, ibuprofen, and pranoprofen were used as model drugs and tween and poloxamer surfactants were tested [60]. The mean particle size of prepared SLNs ranged from 100 to 150 nm. It was found that the drug loading capacity was improved with the most lipophilic drug and a low concentration of the surfactant [60]. Despite some setbacks, lipid nanoparticles continue to be of great interest in the fascinating area of drug delivery technology [61–63].

3
Hydrogels and Networks

Hydrogels are highly swollen, hydrophilic polymer networks that can adsorb large amounts of water and drastically increase in volume. It is well known that the physicochemical properties of the hydrogel depend not only on the molecular structure, the gel structure, and the degree of cross-linking but also on the content and state of water in the hydrogel. Hydrogels have been widely used in controlled release systems [64–66].

Hydrogels that swell and contract in response to external pH [67, 68] are being explored. The pH sensitive hydrogels have a potential use in the site-specific delivery of drugs to specific regions of the GI tract and have been prepared for low molecular weight and protein drug delivery [69]. It is known that the release of drugs from the hydrogels depends on their structure or their chemical properties in response to the environmental pH [10]. These polymers, in certain cases, are expected to reside in the body for a long time and respond to local environmental stimuli to modulate drug release [70]. On the other hand, it is some times expected that the polymers are biodegradable to obtain a desirable device to control drug release [71]. Thus, to be able to design hydrogels for particular applications, it is important to know the variations of systems in their environmental conditions to design them appropriately. Some recent advances in controlled release formulations using hydrogels and networks are discussed here.

3.1
Cross-linked Poly(ethylene Glycol) Networks

The biocompatibility of PEG makes it the polymer of choice for numerous biomedical applications [72, 73]. Graham and coworkers [74–76] pioneered the field of PEG networks cross-linked by diidocyanates as reservoirs for drug delivery. They explored the loading of low molecular weight compounds into and their release from PEG hydrogels. Recently, PEG networks were proposed as reservoirs for delivery of macromolecules, such as proteins, via a transdermal route [77]. PEGs are cross-linked by tris(6-isocyanatohexyl) isocyanurate via urethane/allophanate bond formation to obtain polymeric networks capable of swelling in PBS or ethanol, resulting in gels. The swelling of the networks in PBS and ethanol is governed by the parameters of the initial mixture of PEG and isocyanate, such as molecular weight of the PEG and the ratio of equivalents of isocyano and hydroxy groups. Protein loading into the gels from ethanol is enhanced by the formation of hydrophobic complexes of ion-paired proteins and sodium dodecyl sulfate. Proteins and ethanol release from PEG gels through phospholipid-impregnated membranes mimics that from the biphasic transdermal systems. The spectroscopic data and retention of enzymatic activity of the released proteins indicate that they remain in their native state upon release [77].

3.2
Poly(ε-caprolactone)/poly(ethylene Glycol) Macromer

Drug release from biodegradable or bioerodible polymer matrices has been extensively investigated [78]. The most thoroughly investigated and used bioerodible polymers are the poly(α-hydroxy esters), viz., PLA, PGA, and poly(LA-co-GA) that would degrade into naturally occurring substances [79]. Also, polyanhydrides [80], poly(ortho esters) [81], and poly(α-amino acids) [82] have been developed. Implantable delivery systems using biodegradable polymers are being explored for peptides drugs [83], anticancer therapy [84], hormone therapy [85], antihypertensive drugs [86], and anesthesia [87].

Recently, PEG macromers terminated with acrylate groups and SIPNs composed of PCL and PEG macromer were synthesized and characterized with the aim of obtaining a bioerodible hydrogel that could be used to release tetracycline-HCl for local antibiotic therapy [88]. Polymerization of the PEG macromer resulted in the formation of cross-linked gels due to the multifunctionality of macromer. Non-cross-linked PCL chains interpenetrate into the cross-linked three-dimensional network of PEG. Glass transition temperature (T_g) and melting temperature (T_m) of PCL in the SIPNs were shifted indicating interpenetration of PCL and PEG chains. It was found that the water content increased with increasing PEG weight fraction due to the hydrophilicity of PEG. The weight fraction of PEG in the PCL/PEG SIPNs, the concentration of PEG macromer in the SIPNs preparation, and the nature of PEG might alter the drug release rates

[88]. These studies suggest the hydrophilic nature of PEG that increases the accessibility of water to the polymeric matrix. Also, PCL has been known to degrade very slowly because of its hydrophobic structure that does not allow fast water penetration [88]. PCL degradation by random hydrolytic chain scission of the ester linkages was documented by Pitt et al. [89].

3.3
Gelatin-Polyacrylamide

Ramaraj and Radhakrishnan [90], prepared an interpenetrating hydrogel network from gelatin and polyacrylamide by cross-linking. The swelling behavior of the interpenetrating polymer network system was analyzed in water and in citric acid-phosphate buffer solution at various pHs. The effect of temperature on the swelling behavior of the gels has been analyzed by variations from 25 to 60 °C at physiological pH. The drug release behavior of the gels was also analyzed with temperature variations at physiological pH. An increase in temperature from 25 to 37 °C resulted in a higher and faster drug release [90], which might be due to the extensive swelling and chain relaxation. An increase in temperature beyond 37 °C showed a decrease in drug release followed by erratic changes. At physiological pH, the increase in temperature has accelerated the hydrolysis of acrylamide groups. The polymer matrix, having both acid and amide groups, might possibly experience interactions between them, leading to complex structures through hydrogen bonding. Such a tight structure of the complex restricts the mobility of the polymer segments [91] resulting in slow release of the drug beyond 37 °C.

3.4
Hydroxypropyl Cellulose Gels

Cellulose ethers are common components of pharmaceutical preparations, whether for topical use [92] or oral administration [93, 94]. In solid and semisolid dosage forms, the rate of diffusion of the drug through the gel formed on hydration of the polymer is typically the the key factor determining the release rate [93, 95]. As a result, the effects of formulation variables on drug diffusion rate are of considerable practical relevance [94, 96]. Recently, Alvarez-Lorenzo et al. carried out investigations focusing on the influence of the rheological properties of HPC gels on the in vitro release of theophylline [97]. They performed the experiments with six HPC varieties (mean molecular weight between 5×10^5 and 1.2×10^6, nominal viscosity between 100 and 4000 mPa) at concentrations of 0–2 % (w/w). The theophylline diffusion coefficient declined exponentially with HPC concentration in the case of the lowest-molecular weight HPC, however, the diffusion coefficient remained constant to HPC concentrations of up to 0.8 %, probably because of the high entanglement concentration of the HPC.

3.5
Thermally Reversible Xyloglucan Gels

Materials that exhibit a sol to gel transition in aqueous solution at temperatures between ambient and body temperature are of interest in the development of sustained release vehicles with in situ gelation properties. A compound that has received considerable attention is the polyoxyethylene/polyoxypropylene/polyoxyethylene triblock copolymer Pluronic F127 (poloxomer 407), the thermoreversible gelation was demonstrated by Schmolka [98]. Gels of Pluronic F127 have been explored for application in ophthalmic [99], topical [100], rectal [101, 102], nasal [103], subcutaneous [104], and intraperitoneal [105] administration. However, there are inherent problems associated with triblock copolymers of polyoxyethylene and polyoxypropylene, commercial samples are subject to batch-to-batch variability [106], and laboratory synthesis is complicated by the so-called transfer reaction, which results in the presence of a diblock impurity [107]. These problems may be avoided through the use of block copolymers in which oxybutylene is substituted for oxypropylene as the hydrophobe, which can be tailor made to have the necessary sol-gel transition between ambient and body temperatures to confer in situ gelation characteristics [108]. Yuguchi et al. [109], suggested the polysaccharide xyloglucan, which also exhibits a sol to gel transition in the required temperature region, and which has the additional advantage of recognized non-toxicity and lower gelation concentration, as an alternative polymer.

The xyloglucan polysaccharide derived from tamarind seeds is composed of a [1–4]-β-D-glucan backbone chain, which has [1–6]-α-D-xylose branches that are partially substituted by [1–2]-β-D-galactoxylase. The tamarind seed xyloglucan is composed of three units of xyloglucan oligomers with hexasaccharides, octasaccharides, and nonasaccharides, which differ in the number of galactose side chains. When xyloglucan derived from tamarind seed is practically degraded by β-galactosidase, the resultant product exhibits thermally reversible gelation, the sol-gel transition temperature varying with the degree of galactose elimination [109]. Such gelation does not occur with native xyloglucan. The potential application of xyloglucan gels for rectal [110] and intraperitoneal [111] drug delivery has been reported.

Recently, sustained release vehicles of gels formed in situ following the oral administration of dilute aqueous solutions of a xyloglucan has been assessed by in vitro and in vivo studies [112]. Aqueous solutions of xyloglucan that had been partially degraded by β-galactosidase to eliminate 44 % of the galactose residues, formed rigid gels at concentration of 1.0 and 1.5 % w/w at 37 °C according to Kawasaki et al. [112]. The in vitro release of indomethacin and diltiazem from the enzyme-degraded xyloglucan gels followed root-time kinetics over a period of 5 h at 37 °C at pH 6.8. Plasma concentrations of indomethacin and diltiazem, after oral administration to rats of chilled 1 % w/w aqueous solutions of the enzyme degraded xyloglucan containing the dissolved drug, and a suspension of indomethacin of the same concentration were compared. Constant indomethacin plasma concentrations were noted from both formulations after 2 h and were maintained over a period of at least 7 h. Bioavailability of indomethacin from xyloglucan gels formed in situ was increased approximately three-fold compared with that from the suspension [112]. From these studies it appears that enzyme-degraded xyloglucan gels can be used as prominent vehicles for the oral delivery of drugs.

3.6
Novel Star-Shaped Gel Polymers

Divinyl cross-linking reagents have been often employed for the preparation of star-shaped (*co*)polymers. A linear living polymer was first prepared using a living polymerization technique, and this was subsequently followed by the reaction of its living end with a small amount of a divinyl compound. For instance, the addition of divinylbenzene (DVB) to an anionic living polystyrene (poly[st]) with a central poly(DVB]) gel core was reported [113]. This method was also extended to cationic [114, 115], group transfer [116], and metathesis [117] polymerizations, in which divinyl ethers, divinyl esters, and norbornadiene dimers were used as cross-linkers, respectively. Just adding a bifunctional monomer to a completed living polymerization system could easily carry out this synthetic technique. However, the arm number of the resulting (*co*)polymer could hardly be controlled.

More recently, Ruckenstein and Zhang [118] prepared a novel breakable cross-linker and a pH responsive star-shaped and gel polymer following a traditional method [113–117]. In contrast to the common polymer gels, the star-shaped polymer gel could be easily broken to soluble polymers in an acidic medium. However, it was just swollen in basic or a neutral medium [118]. The hydrolyzed product from the star-shaped polymer was a block copolymer consisting of poly(MMA) and poly(methacrylic acid) segments, and those hydrolyzed from the branched polymers and polymer gels were random copolymers of MMA and methacrylic acid. All the hydrolyzed polymers possessed quite different solubilities than those of their precursors. The authors claimed that these properties might be favorable when used in controlled drug release systems and are relevant to the environment protection.

3.7
Superporous Hydrogels

Porous hydrogels can be prepared by a variety of methods, such as the porosigen technique, the phase separation technique, the cross-linking of individual hydrogel particles, and gas blowing (or foaming) techniques. In the porosigen technique, preparing hydrogels in the presence of dispersed water-soluble porosigens makes porous hydrogels. The major limitation of the phase separation method is that only very limited types of porous hydrogels (such as HEMA and NIPAM) can be prepared, and there is not much control over the porosity of the macroporous hydrogels. Porous hydrogels can also be prepared by cross-linking individual hydrogel particles to form cross-linked aggregates of particles [119]. Pores in such structures are present between hydrogel particles, and the size of pores is smaller that the sizes of the particles. This approach is limited to the absorbent particles that have chemically active functional groups on the surface.

Park and coworkers [120], synthesized porous hydrogels with open channels using the gas blowing [or foaming] technique. The capillary radius of the porous hydrogels are in the range of few hundreds micrometers. Since this pore size is well beyond the pore sizes described by "microporous" (10–100 nm) and "macroporous" (100 nm–10 μm) hydrogels, the hydrogels prepared by these authors were termed as "superporous" hydrogels. Superporous hydrogels prepared by the gas blowing technique were also called hydrogel foams due to the foaming process used in the preparation. Superporous hydrogels were synthesized by cross-linking polymerization of various vinyl monomers in the presence of gas bubbles formed by the chemical reaction of acid and $NaHCO_3$. The polymerization process was optimized to capture the gas bubbles inside the synthesized hydrogels. The use of the $NaHCO_3$/acid system allowed easy control of timing for gelation and foam formation. PF127 was found to be the best foam stabilizer for most of the monomer systems used in their studies [120]. Scanning electron microscope pictures showed interconnected pores forming capillary channels. The capillary channels, which were critical for the fast swelling, were preserved during drying by dehydrating water-swollen hydrogels with ethanol before drying. The ethanol-dehydrated superporous hydrogels reached equilibrium swelling within minutes. These authors have also reported that the equilibrium swelling time could be reduced to less than a minute using a wetting agent. In the present case, residual moisture was used as a wetting agent, since the amount of the moisture content in the dried hydrogels could be easily controlled. Preparation of superporous hydrogels by using the right blowing system, foam stabilizer, drying method, and wetting agent makes it possible to reduce the swelling time to less than a minute regardless the size of the dried gels. The superporous hydrogels can be used where fast swelling and superabsorbent properties are critical, especially in controlled drug release formulations [120].

3.8
Polyelectrolyte Complex Gel of Chitosan and κ-Carrageenan

According to literature, the interest in investigating chitin, chitosan, and their derivatives for use in biology and medicine is rapidly increasing. Chitosan, the deacetylated form of chitin is non-toxic and easily bioabsorbable [121] with gel forming ability at low pH. Moreover, chitosan has antacid and antiulcer activities that prevent or weaken drug irritation in the stomach [122]. Also, chitosan matrix formulations appear to float and gradually swell in acid medium. All these interesting properties of chitosan made this polymer an ideal candidate for controlled drug release formulations [123]. Recent information on chitosan drug delivery systems (viz., nanoparticles, hydrogels, microcapsules, transdermal devices, etc.) is presented in different subsections of the article.

When two oppositely charged polyelectrolytes are mixed in an aqueous solution, a polyelectrolyte complex [124] is formed by the electrostatic attraction between the polyelectrolytes. Sakiyama et al. [125] reported a polyelectrolyte complex gel in cylindrical shape prepared from natural polysaccharide. κ-Carrageenan, which has sulfonate groups, is used as a polyanion component, and chitosan, which has amino groups, is used as polyanion component, and chitosan, which has protonated amino groups, is used as polycation component to afford a strong acid-weak base polyelectrolyte complex. The complex gel swelled in an isotropic manner at ambient pH 10–12, and the swelling maxima was observed in an NaOH solution of pH 10.5. Thus, the swelling of the complex gel was revealed to be sensitive to a rather narrow range of pH. The equilibrium-swelling ratio was also affected by the kind of alkali used. However, in the presence of 4 or 6 % NaCl, the complex gel contracted at any pH [125]. No reports are available on drug release studies.

3.9
Chitosan and Polyether Gel

Yao et al. [126] reported a procedure for the preparation of a semi-interpenetrating hydrogel based on cross-linked chitosan with glutaraldehyde interpenetrating the polyether network. The pH-sensitivity, swelling and release kinetics, and structural changes of the gel in different pH solutions have been investigated [127–129]. It is well known that the physicochemical properties of the hydrogels depend not only on the molecular structure, the gel structure, and the degree of cross-linking but also on the content and state of water in the hydrogel. Since the inclusion of water significantly effects the performance of hydrogels, a study on the physical state of water in the hydrogels is of great importance because it offers useful information on the microstructure and enables us to understand the nature of the interactions between absorbed water and polymers. The dynamic water absorption characteristics, the nature of the correlation between water and the swelling kinetics of chitosan-polyether hydrogels have been studied by applying some novel techniques like positron annihilation life-time spectroscopy and also by widely used techniques like DSC [130].

The effect of the ionic strength of the solution on the hydrolysis rate of the gel has been studied. Rapid hydrolysis of the gel was observed with a decrease in ionic strength of the solution, i.e., a higher degree of swelling with a lower ionic strength of the solution [128]. It was concluded that the hydrolysis of the gel could be controlled by the amount of the cross-linker added. In their further studies, the effect of cross-linker on the swelling behavior of the gel has been studied [128].

Chlorhexidini acetas and cimetidine were used as model drugs for drug release studies. Faster swelling of the gels resulting in more drug release at pH<6 in comparison to the pH >6 was observed [126, 127].

3.10
β-Chitin and Poly(ethylene Glycol) Macromer

Semi-IPN polymer network hydrogels composed of β-chitin and PEG macromer were synthesized for biomedical applications [131]. The thermal and mechanical properties of the hydrogel have been studied. The authors claimed that the tensile strengths of semi-IPNs in the swollen state between 1.35 and 2.41 MPa, are the highest reported values to date for cross-linked hydrogels. The hydrogels have been used as wound covering materials and also studied for their drug release behavior using silver sulfadiazene as a model drug [132].

3.11
β-Chitosan and Poly(ethylene Glycol) Macromer

In their further studies on chitosan for biomedical applications, Lee et al. [133] reported a procedure for preparing semi-IPN polymer network hydrogels composed of β-chitosan and PEG diacrylate macromer, by following a similar procedure to that discussed above. The crystallinity as well as thermal and mechanical properties of gels were reported [133]. Reports on the drug release behavior of the gels are not available.

3.12
Chitosan-Amine Oxide Gel

A procedure for preparing a homogenous chitosan-amine oxide gel was reported [134]. The swelling behavior and release characteristics of the gel were studied in buffer solution (pH 7.4) at room temperature [135–137]. Homogeneous erosion of the matrix and nearly zero-order release of ampicillin trihydrate were observed [136]. The thermal properties of chitosan-amine oxide gel were also reported in subsequent studies [138].

3.13
Poly(ethylene Glycol)-co-poly(lactone) Diacrylate Macromer and β-Chitin

The synthesis and properties of poly(ester-ether-ester) block copolymers based on various lactones and PEG or PPG have been reported in recent years [139, 140]. These polymers are generally used for biomedical materials, such as controlled release of drugs, bioabsorbable surgical sutures, and wound covering materials. Among these polymers, copolymers of L-lactide, D,L-lactide, ε-caprolactone, and PEG have been reported by many workers. Such types of block copolymers have been obtained in bulk by a ring-opening polymerization mechanism. Poly(ester-ether-ester) triblock copolymers composed of PEG and lactones, D,L-lactide, or ε-caprolactone, were cross-linked with β-chitin to prepare SIPN hydrogels by a UV irradiation method [141]. Triblock copolymers were synthesized by bulk polymerization using the low toxic stannous octoate as cat-

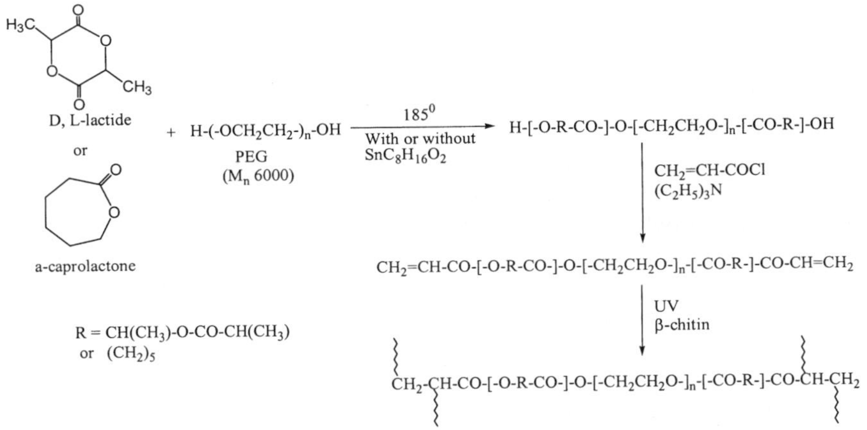

Fig. 3. Schematic representation of PEGLM or PEGCM/β-chitin SIPNs

alyst, or without catalyst. Photo-cross-linked hydrogels exhibited an equilibrium water content in the range of 60–77 %. From DSC analysis, all the hydrogels revealed drastic decreases in crystallinity after photo-cross-linking. In the swollen state, the tensile strength of the semi-IPN hydrogels ranked above 1 MPa. In addition, in spite of their relatively high mechanical strength, elongation at break of swollen hydrogels ranged between 30 and 70 %. Fig. 3 shows the synthetic route to semi-IPNs composed of PEGLM or PEGCM and β-chitin. Vitamin A, vitamin E, and riboflavin were used as model drugs [142]. Recently, Cho et al. [143] reported on a PEG-*co*-poly(lactone) diacrylate macromer and chitosan hydrogels by adopting a similar procedure.

3.14
Chitosan/Gelatin Hybrid Polymer Network

Yao et al. [144] reported a novel hydrogel based on cross-linked chitosan/gelatin with a glutaraldehyde hybrid polymer network, by following similar procedures for polyether and chitosan [126]. The pH dependent swelling behavior and drug release performance of the polymer networks were studied. A drastic swelling behavior of the gels in acidic pH in comparison to basic solution was observed.

Levamisole, cimetidine, and chloramphenicol were used as model drugs. A comparative study on the dependence on the pH value of the release of cimetidine, levamisole, and chloramphenicol from the gel was reported [144]. The results reveal that the drug delivery is controlled by diffusion and relaxation processes, while the diffusion coefficient and relaxation time are highly dependent on the pH of the medium. Moreover, the drug solubility in water obviously has an influence on the release [144].

3.15
Chitosan and D,L-Lactic Acid

Graft copolymerization is one of the best methods to bring together synthetic and natural polymers in order to retain the good properties of natural polymers such as biodegradation and bioactivity. Qu et al. [145] used lactic acid and water-soluble chitosan with a DD of 88 % to synthesize the graft copolymers with hydrophobic synthetic side-chains and hydrophilic, natural main-chains by direct polycondensation without using a catalyst. The formation of hydrogels is explained by the interactions of the hydrophobic polyester side-chains serving as pseudo-cross-links, which stabilize the hydrogel-forming molecules against permanent deformation in the buffer. The specific solution content of hydrogels decreased when the pH value of the buffers was increased, and this change of swellability is reversible. These pH-sensitive hydrogels have potential use in biomedical applications, such as controlled-release systems [145].

3.16
Monolithic Gels

Diffusion through polymers is influenced not only by the polymer's structure, but also by the molecular structure of the solute. The solute structure may also influence the rate of partitioning into the elution medium. However, there is only a low relationship between these factors. The influence of gel structure on the diffusion characteristics of solutes through polyHEMA hydrogels has been reported [146].

The diffusion mechanism was found to be influenced by the nature of the water within the gel and the average pore size of the network. Sorption of solutes was reported to have marked effect on the network structure [147]. Wood et al. [148], investigated the in vitro and in vivo release kinetics of some structurally related benzoic acids from both monolithic and laminated polyHEMA gels. The influence of the physical structure of the polymer network, the stability, concentration, and molecular weight of the solute and the presence of a rate-controlling barrier at the surface of the matrix have been investigated [148]. Zero-order rates of release were achieved by lamination of a rate-controlling barrier to the polymer, and the release rate was modified through changes in the cross-linking density of the barrier layer.

4
Niosomes in Drug Delivery

The self-assembly of non-ionic surfactants into vesicles was first reported in the 1970s by researchers in the cosmetic industry. Since then, a number of groups worldwide have studied non-ionic surfactant vesicles (niosomes) with a view to evaluating their potential as drug carriers. Niosomes are formed from the self-assembly of non-ionic amphiphiles in aqueous media resulting in closed bilayer

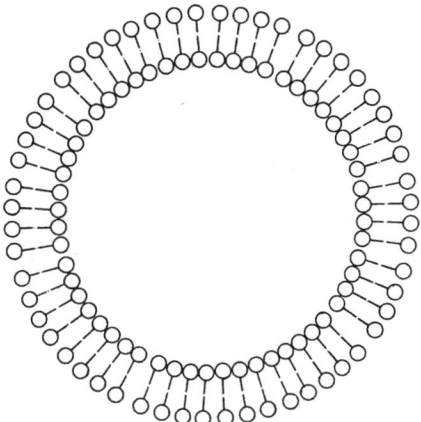

Fig. 4. Schematic representation of a niosome O=hydrophilic head; – = hydrophobic tail

structures (Fig. 4). The assembly into closed bilayers is rarely spontaneous [149] and usually involves some input of energy such as physical agitation or heat. The result is an assembly in which the hydrophobic parts of the molecule are shielded from the aqueous solvent and the hydrophobic head groups enjoy maximum contact with it. These structures are analogous solutes and serve as drug carriers. The low cost, greater stability, and resultant ease of storage of non-ionic surfactants [150] has led to the exploitation of these compounds as alternatives to phospholipids. In a recent review, Uchegbu and Vyas [151] discussed niosome preparation, toxicology studies, and specialized systems.

4.1
Vesicle in Water in Oil Systems

Span surfactant niosomes have been dispersed in oil-in-water emulsions to yield a vesicle in a water-in-oil system, v/w/o, using the same surfactant that was used to make niosomes [152]. The release of CF from these systems followed the trend v/w/o < water in oil (w/o) emulsions < niosome dispersions. The difference between the v/w/o and w/o formulations was minimal. The release of CF encapsulated within these niosomes was influenced by the emulsion oil following the trend, isopropyl myristate > octane > hexadecane and by the nature of the surfactant, following the trend span 20 > span 40 > span 60. Span 80 v/w/o systems had a rather faster release rate due to the unsaturation in the oleyl alkyl chain, which leads to the formation of a more leaky membrane.

Span 60 was found to cause the formation of a gel in the oil phase, which was attributed to the crystallization of span 60 within the oil phase. The net result is an extremely slow release rate from the span 60 v/w/o formulation [152]. These gelled span 60 systems may be stabilized by incorporation of polysorbate 20

[153] and the resultant span 60 v/w/o organogels (oil phase = hexadecane) were found to have a temperature dependent release profile when CF was encapsulated within the niosomes. The release rate was highest at 37 °C when the gel microstructure showed the presence of "tubules" – presumably aqueous water channels along which CF is transported – and slowest at 60 °C when the gel transforms to a recognizable v/w/o system. At this elevated temperature the water channels present in the gel state transform to water droplets within which niosomes are contained. The slow rate of CF transport at 60 °C was presumed to be due to the presence of the oil phase completely surrounding the water droplet through which CF must traverse.

An explanation for this gel formation is sought in the phase transition behavior of span 60. At the elevated temperature (60 °C) which exceeds the span 60 membrane phase transition temperature (50 °C) [154], it is assumed that span 60 surfactant molecules are self-assembled to form a liquid crystal phase. The liquid crystal phase stabilizes the water droplets within the oil. However, below the phase transition temperature the gel phase persists and it is likely that the monolayer stabilizing the water collapses and span 60 precipitates within the oil. The span 60 precipitate thus immobilizes the liquid oil to form a gel. Water channels are subsequently formed when the w/o droplets collapse. This explanation is plausible as the aqueous volume marker CF was identified within these elongated water channels and non-spherical aqueous droplets were formed within the gel [153]. These v/w/o systems have been further evaluated as immunological adjuvants.

4.2
Niosomes in Hydroxypropyl Methyl Cellulose

A transdermal flurbiprofen formulation has been prepared from flubiprofen span 60, cholesterol, DCP (46:50:4) niosomes incorporated within an HPMC semi-solid base containing 10 % glycerin [155]. The in vitro characterization of the formulation is not given, however; this formulation was evaluated in a rat inflammation model.

4.3
Discomes

The solubilization of $C_{16}G_2$ niosomes by solulan C24 results in the formation of the discome phase [156]. This phase consists of giant vesicles of 60 μm in diameter, which encapsulate aqueous solutes such as CF. These large vesicles were found to be of two types: large vesicles that appear ellipsoid in shape and large vesicles that are truly discoid [157]. These morphologies were confirmed by confocal laser scanning microscopy and are only formed in a very specific region of the $C_{16}G_2$ ternary phase diagram (Fig. 5) [157], namely regions 3 and 4. The discomes found in region 3 consist of small spherical, helical, and tubular niosomes (2–10 μm) which are found in a neighboring region,-region 2, while discomes

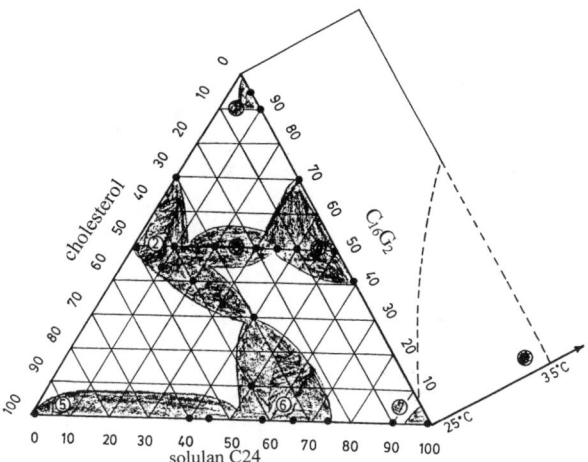

Fig. 5. The hexadecyl diglycerol ($c_{16}G_2$)-cholesterol-solulan C24 ternary phase diagram

found in region 4 do not co-exist with small spherical, helical, and tubular vesicles [158]. On heating the discome dispersion identified in region 3, the large discomes are seen to disappear leaving only the spherical/helical and tubular structures [159]. This is a reversible process and the discomes reform on cooling although the encapsulated aqueous solute is lost to the disperse phase by the heating and cooling cycle (Fig. 6). Once the discomes are destroyed, the release of CF is slowed and represents the release from the remaining small spherical, tubular and helical niosomes.

It is proposed that the system described by region 3 (Fig. 5) may prove useful in ophthalmic delivery as the initial instillation of the formulation into the eye would result in the slow destruction of the discomes and release of an initial burst dose in accordance with the kinetics shown in Fig. 6. The remaining small spherical, helical, and tubular vesicles would then release the rest of the dose slowly to the eye. The large size of the discomes means that clearance from the eye would be slowed down and the destruction of the discomes at 37 °C would result in the release of the encapsulated contents taking place over several minutes (Fig. 6), which would in theory allow the dose to enjoy an increased residence time within the eye. On heating (>35 °C) the discomes formed in region 4 (Fig. 5), a clear isotropic solution is obtained, thought to consist of mixed micelles. Hydrophobic drugs such as paclitaxel may be solubilized by this system and no precipitation of the drug was observed on heating the formulation above 35 °C [157]. This paclitaxel formulation could be stored freeze-dried.

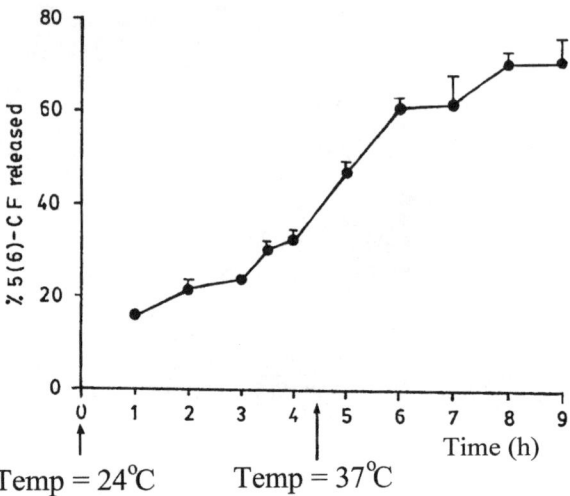

Fig. 6. The release of 5(6) CF from discomes prepared from C16G2, cholesterol, solulan C24 (50:35:15)

4.4
Polyhedral Niosomes

Polyhedral niosomes [160] are formed in low cholesterol regions of the $C_{16}G_2$, cholesterol, solulan C24 ternary phase diagram (Fig. 5). Polyhedral niosomes also encapsulate aqueous solutes such as CF. The vesicles membrane is in the gel phase (Lα) [160], meaning that the hydrocarbon chains enjoy minimum mobility. This gives the vesicles the unusual angular shape. On heating these vesicles above the phase transition temperature (43 °C), the angular shape is not lost and a spherical morphology is observed [161], which on cooling results in an altered morphology. It appears that the heating and cooling cycle causes irreversible changes to the membrane.

Polyhedral niosomes were found to be thermoresponsive Fig. 7 (*a*). Above 35 °C, there was an increase in the release of CF from these niosomes even though the polyhedral shape was preserved until these vesicles were heated to 50 °C. Solulan C24-free polyhedral niosomes do not exhibit this thermoresponsive behavior [160] due to a decrease in the interaction of the polyoxyethylene compound solulan C24 with water at this temperature (due to decreased hydrogen bonding) as identified by viscometry [161]. This observed thermoresponsive behavior was used to design a reversible thermoresponsive controlled release system Fig. 7 (*b*). Thermoresponsive liposomal systems which rely on the changing membrane permeability, when the system transfers from the gel state (Lα) to the liquid crystal state (Lβ) [162], are not reversible. This is not unex-

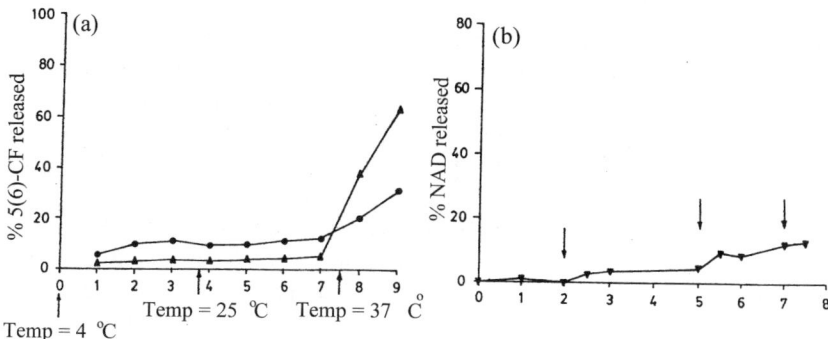

Fig. 7. (*a*) The release of CF from exhaustively dialysed polyhedryl niosomes. ●=$C_{16}G_2$, solulan C24 (91:9); ▲=$C_{16}G_2$, solulan C24 (95:5); (*b*) The release of nicotinamide dinucleotide from polyhedral niosomes prepared from $C_{16}G_2$, solulan (91:9) at 24 °C. Arrow points the time at which the temperature is raised to 37 °C for 10 min

pected, as there is a definite alteration of the membrane characteristics on proceeding through a cooling and heating cycle across the phase transition temperature [161]. Thermoresponsive polyhedral dispersions might find use in dermatology due to their high viscosity [161] and also due to the fact that at 30 °C they are non-thermoresponsive. Thus, they are capable of releasing the encapsulated contents when the ambient temperature is increased to 35 °C or the skin temperature is raised.

5
Microcapsules and Microspheres

The term "microcapsule" is defined, as a spherical particle with the size varying between 50 nm and 2 mm containing a core substance. Microspheres are, in a strict sense, spherically empty particles. However, the terms microcapsules and microspheres are often used synonymously. In addition, some related terms are used as well. For example, "microbeads" and "beads" are used alternatively. Spheres and spherical particles are also employed for a large size and rigid morphology. Due to attractive properties and wider applications of microcapsules and microspheres, a survey of their applications in controlled drug release formulations is appropriate.

5.1
Multiporous beads of Chitosan

Several researchers [163–165] have studied simple coacervation of chitosan in the production of chitosan beads. In general, chitosan is dissolved in aqueous acetic acid or formic acid. Using a compressed air nozzle, this solution is blown into an NaOH, NaOH-methanol, or diaminoethane solution to form coacervate drops. The drops are then filtered and washed with hot and cold water successively. Varying the exclusion rate of the chitosan solution or the nozzle diameter can control the diameter of the droplets. The porosity and strength of the beads correspond to the concentration of the chitosan-acid solution, the degree of N-deacetylation of chitosan, and the type and concentration of coacervation agents used.

The chitosan beads described above have been applied in various fields viz., enzymatic immobilization, chromatography support, adsorbents of metal ions, or lipoprotein, and cell cultures. It was confirmed that the porous surfaces of chitosan beads make a good cell culture carrier. Hayashi and Ikada [166] immobilized protease onto the porous chitosan beads which carry active groups with a spacer and found that the immobilized protease had higher pH, thermal storage stability, and gave rather higher activity towards the small ester substrate, N-benzoyl-L-arginine ethyl ester. In addition, Nishimura et al. [163] investigated the possibilities of using chitosan beads as a cancer chemotherapeutic carrier for adriamycin. Recently, Sharma et al. [167–169] prepared chitosan microbeads for oral sustained delivery of nefedipine, ampicillin, and various steroids by adding the latter to chitosan and then going through a simple coacervation process. These coacervate beads can be hardened by cross-linking with glutaraldehyde or epoxychloropropane to produce microcapsules containing rotundine [170]. The release profiles of the drugs from all these chitosan delivery systems were monitored and showed in general the higher release rates at pH 1–2 than that at pH 7.2–7.4. The effect of the amount of drug loaded, the molecular weight of chitosan, and the cross-linking agent on the drug delivery profiles have been discussed [167–170].

5.2
Coated Alginate Microspheres

Many of the present controlled release devices for in vivo delivery of macromolecular drugs involve elaborate preparation, often employing either harsh chemicals, such as organic solvents [171] or extreme conditions, such as elevated temperature [172]. The conditions have the potential to destroy the activity of sensitive macromolecular drugs, such as proteins or polypeptides. In addition, many devices require surgical implantation and, in some cases, the matrix remains behind or must be surgically removed after the drug is exhausted [173].

Wheatley et al. [174] studied a mild alginate/polycation microencapsulation process, as applied to the encapsulation of bioactive macromolecules such as proteins. The protein drugs were suspended in sodium alginate solution and sprayed into 1.3 % buffered calcium chloride to form cross-linked microcapsules, large (up to 90 %) losses of encapsulation species were encountered, and moderate to strong protein-alginate interactions caused poor formation of capsules. As a result, a diffusion-filling technique for calcium alginate microcapsules that were formed by spraying 10 ml of the sodium alginate solution into

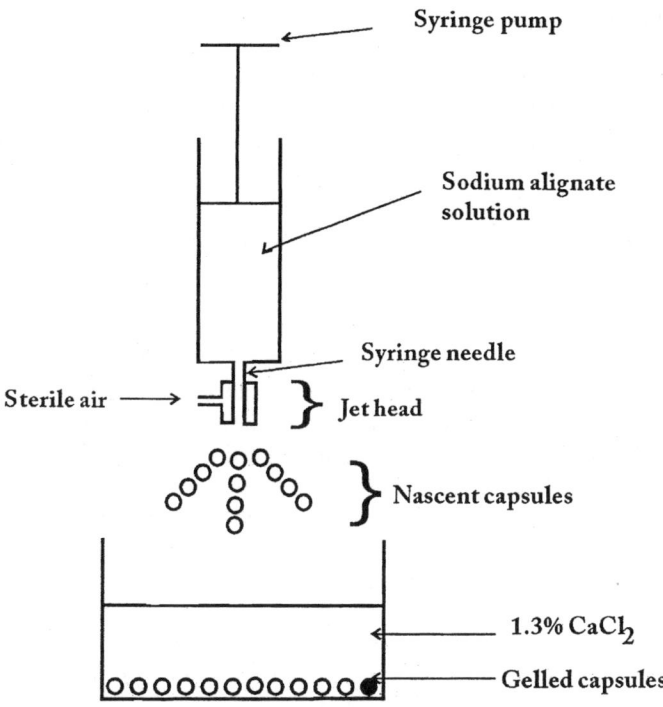

Fig. 8. Spray device for the preparation of calcium alginate microspheres

250 ml of buffered HEPES calcium chloride (13 MM HEPES, 1.3 % CaCl2 pH 7.4) from a 20 ml plastipack syringe through a 22 G needle was developed (Fig. 8).

The protein was then loaded into these capsules by stepwise diffusion from solutions of increasing drug concentration. The drug-loaded capsules were coated with a final layer of polycation. In all, three polycation coatings were used, two prior to filling and one after filling. The first coating strongly influenced the size, integrity, and loading capacity of the capsules. Low concentrations of polycation resulted in poorly formed capsules with very low retention of the drug in the final capsule, while very high concentrations prevented the drug from entering the capsule at the filling stage. The first coat also affected the duration of drug release from the capsule and the size of the burst effect. The second coat had less effect on the capsule integrity, but did influence the drug payload and release profile. The final, sealing-coat had little effect on drug payload and only limited effect on the release profile up to a critical concentration, above which the release profile was not affected. For all coats, increasing polycation concentrations decreased the burst effect, and caused the release profile to be more sustained. Encapsulation of a series of dextrans with increasing molecular weight revealed that the release profile was directly related to the molecular weight of the diffusing species, which was more sustained as the molecular weight increased. Murata et al. [175] investigated alginate gel beads containing chitosan salt. When the bead was placed in bile acid solution it rapidly took bile acid into itself. The uptake amount of taurocholate was about 25 mmol/0.2 g dried gel beads. The phenomenon was observed in the case of the beads incorporating colestyramine instead of chitosan. From the studies reported, it appears that an ion-exchange reaction accompanying the insoluble complex-formation between chitosan salt and bile acid occurs in the alginate gel matrix [175].

5.3
N-(Aminoalkyl).chitosan Microspheres

The most promising encapsulation system yet developed appears to be the encapsulation of calcium alginate beads with PLL. However, the use of this system on a large scale, such as for oral vaccination of animals, is not feasible due to the high cost (\$ 200/g) of PLL. It would therefore be desirable to develop an economic and reliable microencapsulation system based on chitosan and alginate. The better membrane-forming properties of PLL over chitosan were attributed to the following reasons: PLL contains a number of long-chain alkylamino groups that extend from the polyamide backbone. These chains may extend in a number of directions and interact with various alginate molecules, through electrostatic interactions, resulting in a highly cross-linked membrane. Chitosan, on the other hand, has amino groups that are very close to the polysaccharide backbone. Interaction between the charged amino groups of chitosan and carboxylate groups of alginate may be lessened due to steric repulsion between the two molecules.

Goosen and coworkers [176] attempted to mimic the properties of PPL by extending the length of the cationic spacer arm on the chitosan main chain. In the chemical modification, chitosan was first reacted with α-bromoacyl bromide followed by reaction with an amine. The major problem in this procedure was the competing hydrolysis reaction of the bromoacyl bromide. Furthermore, the lack of characterization of the modified chitosan caused ambiguity in the effectiveness of the chitosan modification. No significant difference was found in membrane properties between modified and unmodified chitosan. A two-step synthetic method for attaching long alkylamine side chains to chitosan is represented in Fig. 9.

The approach outlined in Fig. 9 is designed to attach flexible alkylamine side chains to the chitosan polysaccharide backbone, possibly simulating the behavior of PPL. The presence of two amino groups in this side chain may even enhance the membrane-forming properties. Chemical modification of PVA by a similar procedure may also produce an α-polyamine with membrane-forming properties similar to that of PPL [176]. The above synthetic polymer derivatives, as well as chitosan, polyallylamine, and polyethyleneimine, were used to form membrane coatings around the calcium alginate beads in which blue dextran of molecular weight 7.08×10^4 or 26.6×10^4 daltons was entrapped. These microcap-

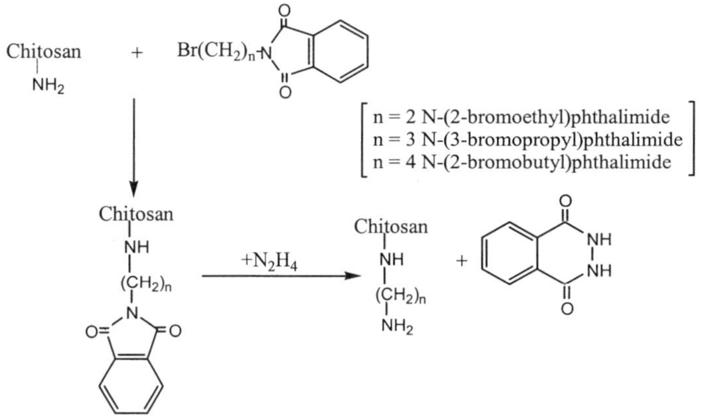

Fig. 9. Modification of chitosan with bromoalkylphthalimides and hydrazine

sules were prepared by extrusion of a solution of blue dextran in sodium alginate into a solution containing calcium chloride and the membrane polymer. Measuring the elution of the blue dextran from the capsules, spectrophotometrically [176], allowed an assessment of membrane integrity and permeability.

5.4
Chitosan/Calcium Alginate Beads

The encapsulation process of chitosan and calcium alginate as applied to the encapsulation of hemoglobin was reported by Huguet et al. [177]. In the first process, the mixture of hemoglobin and sodium alginate is added dropwise to the solution of chitosan and the interior of the capsules thus formed in the presence of $CaCl_2$ is hardened. In the second method, the droplets were directly pulled off in a chitosan-$CaCl_2$ mixture. Both procedures lead to beads containing a high concentration of hemoglobin (more than 90 % of the initial concentration (150 g/l) were retained inside the beads) provided the chitosan concentration is sufficient.

The molecular weight of chitosan (245,000 or 390,000 daltons) and the pH (2, 4, or 5.4) had only a slight effect on the entrapment of hemoglobin, the best retention being obtained with beads prepared at pH 5.4. The release of hemoglobin during the storage of the beads in water was found to be dependent on the molecular weight of chitosan. The best retention during storage in water was obtained with beads prepared with the high molecular weight chitosan solution at pH 2.0. Considering the total loss in hemoglobin during the bead formation and after 1 month of storage in water, the best results were obtained by preparing the beads in an 8 g/l solution of a 390,000 chitosan at pH 4 (less than 7 % of loss with regard to the 150 g/l initial concentration).

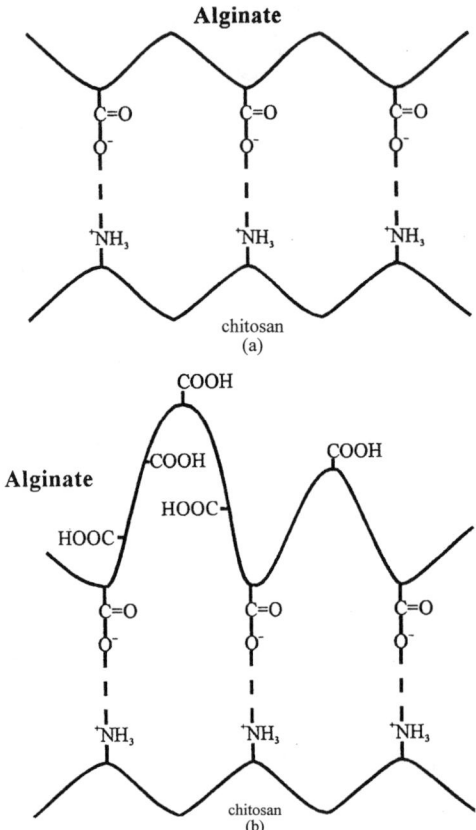

Fig. 10. Schematic representation of ionic interactions between alginate and chitosan at different pH [(a) pH 5.4; (b) pH 2.0]

Similarly, the encapsulation of various molecules (BSA and dextrans with various molecular weights) in calcium alginate beads coated with chitosan has been reported [178, 179]. Their release has been compared and the influence of the conformation, the chemical composition and the molecular weight of the encapsulated materials has been analyzed [178]. The ionic interactions between alginate and chitosan at different pH values are depicted in Fig. 10.

5.5
Poly(adipic Anhydride) Microspheres

In ocular drug delivery, the high rate of tear turnover, and the blinking action of the eyelids lead to short precorneal residence times for applied eye drops. Typically, the washout rate reduces the concentration of the drug in a tear film to one-tenth of its starting value in 4–20 min [180]. As a result, the eye only absorbs a few percent of the administered drug and the duration of the therapeutic action may be quite short. Early reports showed that the formulation of the eye drops is decisive for the availability of an ocular drug [181]. The absorbed amount of the model substance, fluorometholone, and duration in the aqueous humor increased when a suspension was used instead of a solution, and the best result was obtained when the drug was formulated in an ointment. Similar results as with an ointment were obtained when a suspension formulated in a hydrogel was used [182]. The combination of particles with a hydrogel thus increases the bioavailability of an ocular drug. A hydrogel may increase the precorneal residence time of a suspension, and the residence time of the hydrogel, therefore, sets the maximal achievable residence time for a given drug. Water-soluble drugs are generally not retained by hydrogels because of their high diffusion coefficients. One way of solving this problem is to incorporate these drugs into polymeric microparticles.

A novel microsphere-gel formulation was investigated aiming to extend precorneal residence times for ocular drugs [183]. PAA was used for microencapsulation of timolol maleate. A non-aqueous method for the microsphere preparation was employed due to the hydrolytic sensitivity of the polymer. Microspheres were prepared with an average diameter of 40 µm. The polymer and the microspheres were characterized before and during degradation using SEC, DSC, XRD IR, and SEM [183]. The microspheres had a smooth external surface and a hollow center surrounded by a dense outer shell. Degradation of the microspheres resulted in a constant release of adipic acid, the degradation product, indicating a surface-eroding degradation mechanism. The surface erosion of the polymer controlled the release of incorporated substance, timolol maleate. The drug release rate profile appeared to be suitable for ocular drug delivery. However, the initial drug release rate was decreased to some extent when the PAA-microspheres were incorporated into an in situ gelling polysaccharide [183], Gelrite. The authors claimed that the improved ocular bioavailability of these novel microsphere-gel delivery formulations remains to be compared with that of ordinary eye drops.

5.6
Gellan-Gum Beads

Gellan gum is a linear anionic polysaccharide produced by the microorganism *Pseudomonas elodea* [184, 185]. The natural form of the polysaccharide consists of a linear structure with a repeating tetrasaccharide unit of glucose, glucuronic acid, and rhamnose [185–187] in a molar ratio of 2:1:1. Native gellan is partially acylated with acetyl and L-glyceryl groups located on the same glucose residue [188]. XRD analysis shows that gellan gum exists as a half-staggered, parallel, double helix which is stabilized by hydrogen bonds involving the hydroxymethyl groups of one chain and both carboxylate and glyceryl groups of the other [189]. The presence of acetyl or glyceryl groups does not interfere with the double formation but does alter its ion-binding ability. The commercial gellan gum is the deacetylated compound obtained by treatment with alkali [184], yielding the gum in its low acyl form in which the acetate groups do not interfere with helix aggregation during gel formation. Gellan forms gels in the presence of mono and divalent ions, although its affinity for divalent ions is much stronger [190]. Milas et al. [191] showed a mechanism of gelation based on aggregation of the double helix controlled by the thermodynamics of the solution in which the nature of the counter-ion is of prime importance. The apparent viscosity of the gellan gum dispersions can be markedly increased by increasing both pH and cation concentration [192, 193]. Gellan gum is mainly used as a stabilizer or thickening agent and it has a wide variety of applications, particularly in the food industry [190, 194], as a bacterial growth media [195, 196], and in plant tissue culture [197]. Its medical and pharmaceutical uses are in the field of sustained release. Due to its characteristic property of temperature-dependent and cation-induced gelation, gellan has been used in the formulation of eye drops, which gellify on interaction with the sodium ions naturally present in the eye fluids [197–199].

Microcapsules containing oil and other core materials have been formed by complex coacervation of gellan gum-gelatin mixtures [200]. Deacetylated gellan gum was used to produce a bead formulation containing sulphamethizole by a hot extrusion process into chilled ethyl acetate [193]. Recently, the ability of gellan gum to form gels in the presence of calcium ions was investigated, this enabled capsules to be prepared by gelation of the polysaccharide around a core containing starch [201, 203], or oil [203].

Kedzierewicz et al. [204] adopted a rather simpler method than the ones used so far, i.e., the ionotropic gelation method, to prepare gellan gum beads. Gellan gum beads of propranolol-hydrochloride, a hydrophilic model drug were prepared by solubilizing the drug in a dispersion of gellan gum and then dropping the dispersion into calcium chloride solution. Major formulation and process variables, which might influence the preparation of the beads and drug release from gellan gum beads, were studied. Very high entrapment efficiencies were obtained (92 %) after modifying the pH of both the gellan gum dispersion and the calcium chloride solution. The beads could be stored for 3 weeks in a wet or dried state without modification of the drug release. Oven-dried beads released the drug somewhat more slowly than the wet or freeze-dried beads. The drug release from the oven-dried beads was slightly affected by the pH of the dissolution medium [204]. Gellan gum could be a useful carrier for the encapsulation of fragile drugs and provides new opportunities in the field of bioencapsulation.

5.7
Poly(D, L-lactide-*co*-glycolide) Microspheres

The treatment of infiltrating brain tumors, particularly oligodendrogliomas, requires radiotherapy, which provides a median survival of 3.5–11 years [205]. Since IdUrd is a powerful radio sensitizer [206], the intracranial implantation of

IdUrd loaded microparticles within the tumor might increase the lethal effects of γ-radiation of malignant cells having incorporated IdUrd. The particles can be administered by stereotactic injection, a precise surgical injection technique [207]. This approach requires microparticles of 40–50 μm in size releasing in vivo their content over 6 weeks, the standard period during which a radiotherapy course must be applied.

The solvent evaporation process is commonly used to encapsulate drugs into PLGA microparticles [208]. It is well known that the candidate drugs must be soluble in the organic phase. In the case where the active ingredient is not oil soluble, other alternatives can be considered. The W/O/W-multiple emulsion method is particularly suitable for the encapsulation of highly hydrophilic drugs. For drugs which are slightly water soluble, like IdUrd (2 mg/ml), other approaches must be investigated to achieve significant encapsulation: dissolution of the drug in the organic phase through the use of a cosolvent or dispersion of drug crystals in the dispersed phase. In the latter case, it is often admitted that the suspension of crystals in the organic phase can lead to an initial drug release, which is difficult control [209, 210]. To reduce IdUrd particle size, two-grinding processes were used, spray-drying and planetary ball milling [211–213]. The optimal conditions of grinding were studied through experimental design and the impact on in vitro drug release from PLGA microspheres was then examined. More recently, Geze et al. [214], studied IdUrd loaded PLGA microspheres with a reduced initial burst in the in vitro release profile, by modifying the drug grinding conditions. IdUrd particle size reduction has been performed using spray drying or ball milling. Spray drying significantly reduced drug particle size with a change of the initial crystalline form to an amorphous one and led to a high initial burst. Conversely, ball milling did not affect the initial IdUrd crystallinity. Therefore, the grinding process was optimized to emphasize the initial burst reduction. The first step was to set qualitative parameters such as ball number, and cooling with liquid nitrogen to obtain a mean size reduction and a narrow distribution. In the second step, three parameters including milling speed, drug amount, and time were studied by a response surface analysis. The interrelationship between drug amount and milling speed was the most significant factor. To reduce particle size, moderate speed associated with a sufficient amount of drug (400–500 mg) was used. IdUrd release from microparticles prepared by the O/W emulsion/extraction solvent evaporation process with the lowest crystalline particle size (15.3 μm) was studied to overcome the burst effect. In the first phase of drug release, the burst was 8.7 % for 15.3 μm compared to 19 % for 19.5 μm milled drug particles [214].

In the other procedure, Rojas et al. [215] optimized the encapsulation of BLG within PLGA microparticles prepared by the multiple emulsion solvent evaporation method. The role of the pH of the external phase and the introduction of the surfactant tween 20, in the modulation of the entrapment and release of BLG from microparticles were studied. Better encapsulation of BLG was noticed on decreasing the pH of the external phase. Addition of tween 20 increased the encapsulation efficiency of BLG and considerably reduced the burst release effect.

In addition, tween 20 reduced the number of aqueous channels between the internal aqueous droplets as well as those communicating with the external medium. The inventors claimed that these results constitute a step ahead in the improvement of an existing technology in controlling protein encapsulation and delivery from microspheres prepared by the multiple solvent evaporation method [215].

Blanco-Prieto et al. [216] studied the in vitro release kinetics of peptides from PLGA microspheres, optimizing the test conditions for a given formulation, which is customary to determine the in vitro/in vivo correlation. The somatostatin analogue valpreotide palmoate, an octapeptide, was microencapsulated into PLGA 50:50 by spray drying. The solubility of this peptide and its in vitro release kinetics from the microspheres were studied in various test media. The solubility of valpreotide palmoate was approximately 20–40 µg/ml in 67 mM PBS at pH 7.4, but increased to 500–1000 µg/ml at a pH of 3.5. At low pH, the solubility increased with the buffer concentration (1–66 mM). Very importantly, the proteins BSA solution or human serum appeared to solubilize the peptide palmoate, resulting in solubilities ranging from 900 to 6100 µg/ml. The release rate was also greatly affected by the medium composition. The other results are: in PBS of pH 7.4 only 33±1 % of the peptide was released within 4 days, whereas, 53±2 and 61±0.9 % were released in 1 % BSA solution and serum, respectively. The type of medium was found to be critical for the estimation of the in vivo release. From these investigations, it was concluded that the in vivo release kinetics of valpreotide palmoate from PLGA microspheres in rats were qualitatively in good agreement with those obtained in vitro using serum as release medium, while sterilization by γ-irradiation had only a minor effect on the in vivo pharmacokinetics [216].

5.8
Alginate-Poly-L-lysine Microcapsules

Transplantation of islets of Langerhans as a means of treating insulin-dependent diabetes mellitus has become an important field of interest [217–219]. However, tissue rejection and relapse of the initial autoimmune process have limited the success of this treatment. Immunoisolation of islets in semipermeable microcapsules has been proposed to prevent their immune destruction [220, 221]. Nevertheless, a pericapsular cellular reaction eventually develops around microencapsulated islets, inducing graft failure [222]. Since empty microcapsules elicit a similar reaction [223], the reaction is not related to the presence of islets within the capsule but is, at least partially, caused by the capsule itself. Consequently, microcapsule biocompatibility appears to constitute a major impediment to the successful microencapsulated islet transplantation.

Smaller microcapsules [<350 µm] offer many advantages over standard microcapsules [700 to 1500 µm], including reduced total implant volume, better insulin kinetics [224], improved cell oxygenation [225], and potential access to diverse implantation sites, such as the spleen and liver. A new electrostatic pulse system produces microcapsules of <350 µm in diameter [226, 227] as compared with 700 to 1500 µm produced with the usual air-jet system. Lum et al. [226] have suggested that these smaller microcapsules exhibit a higher degree of biocompatibility compared to standard microcapsules. However, no quantitative data or comparative studies have been published addressing this issue.

Previous investigations on alginate-PLL microcapsule biocompatibility have focused mainly on intraperitoneal transplantation of microcapsulated islets into diabetic rats. The use of peritoneal implantation for biocompatibility studies is hindered by the fact that, in this site, microcapsules are unevenly distributed. Free floating microcapsules, which are easily recovered, are less likely to show pericapsular fibrosis than irretrievable microcapsules. The selection bias hinders a quantitative evaluation of microcapsule biocompatibility. The low recovery of smaller microcapsules increases the selection bias. To overcome these methodological problems, Pariseau et al. [228], developed and carefully validated a method for the in vivo comparative evaluation of microcapsule biocompatibility. This technique comprises implantation of microcapsules into rat epididymal fat pads, retrieval of fat pads after fixed time periods, and histological evaluations with the use of a fibrosis score. The microcapsule recovery rate from this site was 99.6 ± 0.75 % [228]. The pericapsular reaction is uniform within one fat pad and between fat pads, allowing random sampling and comparative studies [228].

Robitaille et al. [229] reported investigations comparing the biocompatibility of microcapsules <350 µm in diameter made with an electrostatic pulse system with that of microcapsules of 1247 ± 120 µm in diameter made with the standard air-jet system with the objectives: (1) to compare the biocompatibility of smaller and standard microcapsules while maintaining either equal implant volume or equal alginate content, and (2) to analyze the biocompatibility of smaller versus standard microcapsules with respect to the total implant surface exposed to the surroundings. To evaluate the biocompatibility, 200, 1000, 1120, 1340, or 3000 smaller microcapsules (<350 µm) or 20 standard microcapsules (1247 ± 120 µm) were implanted into rat epididymal fat pads, retrieved after 2 weeks, and evaluated histologically. The average pericapsular reaction increased with the number of small microcapsules implanted ($P<0.05$; 3000 vs. 200, 300 vs. 1000, and 1000 vs. 200 microcapsules). At equal volume and alginate content, standard microcapsules caused a more intense fibrosis reaction than smaller microcapsules ($P<0.05$). In addition, 20 standard microcapsules elicited a stronger pericapsular reaction ($P<0.05$) although the later represented a 3.4-fold larger total implant surface exposed. Finally, from these investigations it appears that the microcapsules of diameters <350 µm made with an electrostatic pulse system are more biocompatible than standard microcapsules [229].

5.9
Cross-Linked Chitosan Microspheres Coated with Polysaccharide or Lipid

The procedure for the preparation of cross-linked chitosan microspheres coated with polysaccharide or lipid for intelligent drug delivery systems is illustrated in Fig. 11 [230].

The microspheres were prepared with an inverse emulsion of 5-FU or its derivative solution of hydrochloric acid and chitosan in toluene containing span 80. Chitosan was cross-linked through Schiff's salt formation by adding glutaraldehyde in toluene solution. At the same time, the amino derivatives of 5-FU were immobilized, obviously resulting in an increase in the amount of drug within the microspheres. The microspheres were coated with anionic polysaccharides (e.g., carboxymethylchitin, etc.) through a polyion complex formation reaction. In the case of lipid-coated microspheres, the microspheres along with DPPC were dispersed in chloroform. After evaporation of the solvent, microspheres were obtained coated with a DPPC lipid multilayer, which exhibited a transition temperature of a liquid crystal phase at 41.4 °C. The diameter range of microspheres was 250–300 nm with a narrow distribution. The stability of the dispersion was improved by coating the microspheres with an anionic polysaccharide or a lipid multilayer Fig. 12.

A comparative study on the release of 5-FU and its derivatives from polysaccharide-coated microspheres MS [CM] was carried out in physiological saline at 37 °C. Data indicated that the 5-FU-release rate decreased in the order: free-5-FU>carboxymethyl type 5-FU>ester type 5-FU. The results revealed that the coating layers on the microspheres were effective barriers to 5-FU release.

The lipid multilayers with a homogeneous composition generally show a transition of gel-liquid crystal. When the temperature is raised to 42 °C, which is higher than the phase transition of 41.4 °C, the released amount of 5-FU increased, while the amount of drug delivered decreased at 37 °C, which is lower

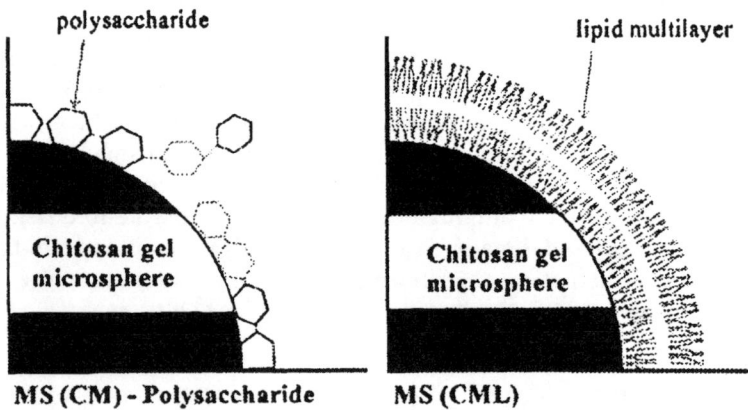

Fig. 11. Proposed structures of chitosan gel microspheres coated with polysaccharide or lipid

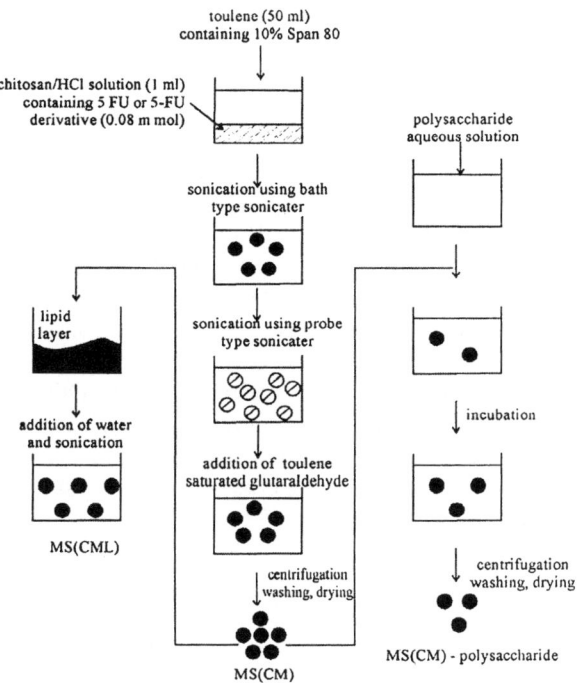

Fig. 12. Preparation process of MS (CM), MS (CML) and MS (CML) and MC (CM) polysaccharide

than the transition temperature. Due to the improved recognition function of polysaccharide chains for animal cell membranes, it is reasonable to develop targeting delivery systems from polysaccharide-coated microspheres, MS [CM].

5.10
Chitosan/Gelatin Microspheres

In their studies on pharmaceutical applications of chitin and chitosan, Yao and coworkers [231] reported on chitosan/gelatin network polymer microspheres for the controlled release of cimetidine. The drug loaded microspheres were prepared by dissolving chitosan, gelatin (1:1 by weight), and cimetidine in 5 % acetic acid. A certain amount of tween 80 and liquid paraffin at a water to oil ratio of 1:10 was added to the chitosan/gelatin mixture under agitation at 650 rpm at 30 °C. A suitable amount of 25 % aqueous glutaraldehyde solution was added to the inverse emulsion and maintained for 2 h. Finally, the liquid paraffin was vaporized under vacuum to obtain microspheres.

The drug release studies were performed in HCl (pH 1.0) and phosphate buffer (pH 7.8) at an ionic strength 0.1 m/l. A pH dependent pulsed-release behavior of the HPN matrix was observed [231]. Moreover, the release rate can be controlled via the composition of the HPN and the degree of deacetylation of chitosan.

5.11
Cross-Linked Chitosan Network Beads with Spacer Groups

A novel technique for the preparation of pH sensitive chitosan beads was reported. DFS, Thy-HCl, CPM, and INH were used as model drugs [232–234]. In these studies, widely used products in medical and pharmaceutical areas, viz., glycine and PEG were employed as spacer groups to enhance the flexibility of the polymer networks and influence the swelling behavior through macromolecular interactions. The procedure is based on adding drugs to chitosan solution followed by bead formation by simple coacervation [235]. The swelling behavior, solubility, hydrolytic degradation, and drug loading capacity of the beads were investigated [234, 236]. The effect of the cross-linker was studied by varying the amount of the cross-linker [234].

The beads exhibit high pH sensitivity. The swelling ratio of the beads at pH 2.0 is obviously higher than that at pH 7.4. This pH sensitive swelling is due to the transition of the bead network between the collapsed and the expanded forms, which is related to the ionization degree of the amino groups on chitosan in different pH solutions. In both these systems, a near zero-order release is observed for about 4 days. The amount and percent release in the chitosan-PEG system is a bit higher when compared to the chitosan-glycine system, due to the water diffusivity and pore forming properties of PEG. The effects of initial drug concentration and the cross-linking agent on the delivery profiles were reported as well [232–235].

5.12
1,5-Diozepan-2-one (DXO) and D,L-Dilactide (D,L-LA) Microspheres

The most successful class of degradable polymers so far has been that of the aliphatic polyesters. The degradation takes place via hydrolysis of the ester linkages in the polymer backbone [237]. These materials must be extensively tested and characterized since many of them are new structures. However, this alone is not sufficient. Equally important is the identification of their degradation mechanisms and degradation products. Since high molecular weight polymers are seldom toxic, the toxicity and tissue response after the initial post operative period is related to the compounds formed during degradation.

The most important polymer on the market today is PLA, which upon degradation yields lactic acid, a natural metabolite in the human body [79]. The formation of natural metabolites should be advantageous as the body has routes to eliminate them. Other commercial degradable materials are polyparadioxane [238], the copolymer of glycolic acid and trimethylene carbonate [239], which does not give natural metabolites when degraded. Their most important characteristic is probably the fact that the degradation products are harmless in the concentrations present.

Albertsson and coworkers [240–244] carried out extensive research to develop polymers in which the polymer properties are altered for different applications. The predominant procedure is ring-opening polymerization which provides a way to achieve pure and well defined structures. They have utilized cyclic monomers such as lactones, anhydrides, carbonates, ether-lactones. The work involved the synthesis of monomers not commercially available, studies of polymerization to form homopolymers, random and block copolymers, development of cross-linked polymers and polymer blends, surface modification in some cases, and characterization of the materials formed. The characterization is carried out with respect to the chemical composition and both chemical and physical structures, the degradation behavior in vitro and in vivo, and in some cases the ability to release drug components from microspheres prepared from the polymers.

Copolymers of DXO and D,L-LA were synthesized in different molar concentrations in their recent attempts [245]. In vitro hydrolysis studies on the copolymers showed degradation times up to 250 days. Microspheres intended for drug delivery were prepared from poly[TMC-co-CL], PAA and PLGA [245]. SEM studies showed that the microsphere morphology depends on the concentration at the time of preparation and on the choice of polymer. The drug release profiles showed dependence on the polymer degradation behavior and on the water penetration into the microsphere [245].

5.13
Triglyceride Lipospheres

Liquid drug-delivery systems, which do not require surgical implantation, may use vehicles made of microparticulates or colloidal carriers composed of lipids, carbohydrates, or synthetic polymer matrices. Liposomes, the most widely studied of these vesicles, can be formulated to include a variety of compositions and structures that are potentially non-toxic, degradable, and non-immunogenic. To produce a long acting local anesthetic effect, vesicles have been used to entrap dibucaine [246], methoxyflurane [247], tetracaine [248], and lidocaine [249] using formulations with polylactic acid, lecithin, iophendylate and phosphatidylcholine, and cholesterol, respectively. With varying degrees of success, these treatments have provided neural blockade for periods far outlasting those produced by any drug given alone.

Masters and Domb [250] reported on an injectable drug delivery system that uses liposomes [251] to release the local anesthetic, bupivacaine, from a liposomal matrix that is both biodegradable and biocompatible to produce SLAB. Bupivacaine due to its minimum vasodilating properties was preferred to other local anesthetics (e.g., lidocaine) allowing the released drug to remain at the site of injection longer [252]. Lipospheres are an aqueous microdispersion of water insoluble, spherical microparticles (0.2 to 100 μm in diameter), each consisting of a solid core of hydrophobic triglycerides and drug particles that are embedded with phospholipids on the surface. The in vivo studies with liposheres have shown that a single bolus injection can deliver antibiotics and anti-inflammatory agents for 3 to 5 days [253] and also, control the delivery of vaccines [254,255]. Recent reports were of a bupivacaine-liposhere formulation, which produced 1–3 days reversible sensory and motor SLAB when applied directly to the rat sciatic nerve [250]. The particle size of the liposheres was in between 5 and 15 μm, with over 90 % surface phospholipid. Lipospheres released bupivacaine over two days under ideal sink conditions. Administration of lipospheress to nerves produced dose-dependent and reversible block. SLAB was observed for 1–3 days in various in vivo tests: (a) Hind paw withdrawal latency to noxious heat was increased over 50 % for a 96 h period after administration of 3.6 % or 5.6 % bupivacaine-lipospheres. The 3.6 % and 5.6 % doses were estimated to release bupivacaine at 200 and 311 mg drug/h, respectively, based on release spanning 72 h. Application of 1.6 % bupivacaine-lipospheres increased withdraw latency by 25–250 % but for only a 24 h duration; (b) similarly, the vocalization threshold to hind paw stimulation was increased by 25–50 % for 72 h following administration of 3.6 % bupivacaine-lipospheres; (c) finally it was concluded that the sensory blockade outlasted or equaled the corresponding motor block durations for all liposphere drug dosages [250].

5.14
Glutamate and TRH Microspheres

L-Glutamate is the principal excitatory neurotransmitter in the mammalian central nervous system [256] and has been shown to stimulate trigeminal motoneurons within the trigeminal motor nucleus in acute, short-term physiological studies [257]. Since trigeminal motoneurons innervate the muscles of mastication and the activity patterns [EMG] of these muscles directly affect the growth/development of the craniofacial skeleton by biomechanical forces produced [258], Byrd and coworkers [259, 260] tried and successfully used gultamate-impregnated microspheres to stimulate trigeminal motoneurons in situ within the brainstem of young rats to produce skeletal alterations. The underlying premise was that increased delivery of glutamate in the proximity of trigeminal motoneurons would increase the activity of both those motoneurons and the masticatory muscles they supply. Indirect physiological evidence for this premise was provided by the presence of (1) more pronounced implant-side wear facets on the mandibular incisors in rats with gultamate-microsphere implants, and (2) deviation of their facial skeletons toward the implant side [259, 260]. The microspheres used were of biodegradable, polyanhydride construction and released gultamate at a controlled rate within the intact rat. The use of biodegradable polyanhydride microspheres as drug-carrier matrices [261] is an effective method for the delivery and long-term release of neuroactive substances to a specific locus within the CNS with little risk of infection [260].

The tripeptide TRH has been confirmed as an important neurotransmitter/neuromodulator within the brain stem region of the CNS [262–266]. Trigeminal motoneurons are highly immunoreactive for TRH [264] and also TRH actually increases the excitability of other brain stem motoneurons in vitro [266]. TRH therefore proved to be useful to increase activity levels of trigeminal motoneurons in vivo. Byrd et al. [259] investigated in the chronic, long-term effects of administering TRH in the proximity of trigeminal motoneurons in vivo, and compared any craniofacial sequelae with those effected by glutamate microspheres.

Byrd et al. [267] investigated the sequelae of sustained, in vivo delivery of two important neurotransmitter substances, glutamate and TRH, upon craniofacial growth and development. In their studies, the relative effects of glutamate and TRH microspheres stereotactically placed in proximity to trigeminal motoneurons within the trigeminal motoneurons were compared. Stereoactive neurosurgery at 35 days was conducted on 4 experimental groups comprising 10 male Sprague-Dawley rats per group. The data were collected after killing 5 rats of each group at 14 and 21 days. Histology revealed that the implants were clustered in the pontine reticular formation, close to ventrolateral tegmental nucleus. Both glutamate and TRH rats had implant-side deviation of their facial skeletons and snout regions. 4×2 ANOVA and post hoc t-tests revealed significant [$p < 0.05, 0.01$] differences between groups and sides for motoneuron count, muscle weight, and osteometric data.

5.15
Polyelectrolyte Complexes of Sodium Alginate Chitosan

Polyelectrolyte complexes (PECs) are formed by the reaction of a polyelectrolyte with an oppositely charged polyelectrolyte in an aqueous solution. Polysaccharides, which have bulky pyranose rings and a highly stereoregular configuration in their rigid, linear backbone chains, have been frequently studied [268]. PECs have numerous applications such as membranes, antistatic coatings, environmental sensors, and chemical detectors, medical prosthetic materials, etc. [269]. Among these, their wide use as membranes for dialysis, ultrafiltration, and other solute separation processes are of special interest and these properties also made their use in microcapsule membranes possible. Microcapsules can be used for mammalian cell culture and the controlled release of drugs, vaccines, antibiotics, and hormones [269–271], To prevent the loss of encapsulated material, the microcapsules should be coated with another polymer that forms a membrane at the bead surface. The most well known and promising system is the encapsulation of alginate beads with PLL. Because this system has a limitation due to the high cost of PLL, other systems such as alginate beads coated with chitosan or its derivatives have been developed [176, 177]. Few results have been reported about the formation of PECs of alginate with chitosan under acidic conditions. Although alginate/chitosan microcapsules have been studied a lot, the studies have been limited to a narrow pH region due to the solubility of chitosan.

Lee et al. [272] described a new procedure to overcome the solubility of chitosan. In this procedure, chitin was heterogeneously deacetylated with a 47 % sodium hydroxide solution and followed by a homogeneous reacetylation with acetic anhydride to control the N-acetyl content of the chitosan having a similar molecular weight. The chitins having different degrees of N-deacetylation were complexed with sodium alginate, and the formation behavior of PECs was examined by viscometry in various pH ranges. The maximum mixing ratio (R_{max}) increased with a decrease in the degree of N-acetylation of chitosan at the same pH, and with a decrease in pH at the same degree of N-acetylation. Similarly, N-acylated chitosans were also prepared. The N-acyl chitosans scarcely affected the formation behavior of PECs with sodium alginates. For the application of the PECs produced, the microencapsulation of a drug was performed and the release property of the drug was tested. The microcapsules were prepared in one step by the extrusion of a solution of a guaifenesin and sodium alginate into a solution containing calcium chloride and chitosan through inter-polymeric ionic interactions. The drug release during the storage of drug-loaded microcapsules in saline dependent on the pH at which the microcapsules were formed and the kind of N-acyl groups introduced to the chitosan. The microcapsules prepared at pH 4.8 showed a minimum release rate and the release rate varied with the pH due to loop formation of the backbone chains of polyelectrolytes. The N-acyl groups introduced to the chitosan enhanced the release rate remarkably.

5.16
Albumin Microspheres

Albumin is an attractive macromolecular carrier and widely used to prepare microspheres and microcapsules, due to its availability in pure form and its biodegradability, non-toxicity, and non-immmunogenicity [273]. A number of studies have shown that albumin accumulates in solid tumors [274, 275] making it a potential macromolecular carrier for the site-directed delivery of antitumor drugs. Recently, Katti and Krishnamurti [276] prepared albumin microspheres by suspension cross-linking in the absence of any surfactant using paraffin oil as the dispersion medium and formaldehyde as the cross-linking agent. They characterized the microspheres by SEM and found them to be spherical having a particle size distribution in the range of 50–400 μm. A preliminary drug release study of chlorothiazide in vitro indicated a diffusion controlled release of the drug. The authors claimed that this method is simple, cost effective, and moreover, promising technique for large-scale manufacture [276].

6
Films and Membranes

The choice of polymers suitable for forming the carrier film matrix and the barrier film are dictated by several factors, including:
- compatibility with the gastric environment,

- polymer stability during the time of complete drug delivery,
- appropriate mechanical properties, i.e., capability of forming self-supporting films when
- containing the drug at high loading,
- no appreciable swelling in water and having a softening point above 37 °C,
- ease of fabrication,
- cost.

Several polymers were found to fit all or most of the above criteria and were used to prepare the carrier films. Many polymers have been used for this purpose, viz., ethyl cellulose, poly(γ-benzyl glutamate), poly(vinyl acetate), cellulose acetate phthalate, and the copolymer of methyl vinyl ether with maleic anhydride. In addition to the base polymers, plasticizers were often needed to impart a suitable degree of flexibility. Plasticizers, which are found to be compatible with polymeric materials include, acetylated monoglycerides, esters of phthalic acid such as dibutyl tartarate, etc. An excipient was usually incorporated into the matrix of the carrier films. The excipients used were water-soluble materials, which are capable of creating channels in the polymer matrix and facilitate diffusion of the drug. PEGs of different molecular weights were used for this purpose.

Generally, the barrier or rate-controlling films are more permeable to water than the carrier films. The materials used for this purpose consisted of a base, film forming water-soluble polymer in combination with at least one hydrophilic component such as hydroxypropyl methylcellulose or polyvinylpyrrolidone. The polymers used were the same as those for the carrier films. Some recent developments are discussed herein.

6.1
Cross-Linked Gelatin Films

Bioadhesive materials that can adhere to soft tissues have a potential in medical applications. Controlled drug delivery through mucous membranes, such as buccal, rectal, and vaginal mucus requires such bioadhesive materials. Some transmucous drug delivery systems have been commercialized using a mixture of hydroxypropyl cellulose and carbopol 934 [277], and carboxy vinyl polymer [278], which are reported to be adhesive to natural tissue. In cases of damage, abdominal, tendon, and pericardial tissues tend to adhere to the surrounding tissues, causing impairment of their function. In these cases, biomaterials used in the prevention of tissue adhesion temporarily cover the wound site until tissue healing is completed. Recently, an anti-adhesion material was commercialized as Seprafilm. It is made of a mixture of hyaluronic acid and carboxymethyl cellulose and it is claimed that it will adhere well to tissue and moist surfaces without suturing [279]. Most of the biomaterials used for these purposes contain many ionizable groups, mainly carboxy groups, which are able to adhere to soft tissues without sutures and adhesives, probably through hydrogen bonding between the carboxy group of the biomaterials, which are swollen with water [280].

More recently, Matsuda et al. [281] reported a procedure for preparing strong bioadhesive gelatin films by introducing free aldehyde groups that can form covalent bonds through Schiff base formation with amino groups of biological soft tissues.

6.2
Transdermal Devices

The transdermal route has been recognized as an interesting and viable technique to provide controlled delivery of drugs to the systemic circulation, since the skin remains intact. Extensive studies have been carried out to provide drugs through this route [282]. TDS are polymeric patches containing dissolved or dispersed drugs that deliver therapeutic agents at a constant rate to the body. The simplest type of such systems contains a drug dissolved in an adhesive that is applied to an impermeable backing membrane. This composite matrix is attached to a release liner. Although this type of delivery device is structurally simple, it places multiple requirements on the adhesive with respect to adhesion, drug storage, and release. Several incidences of crystallization of drugs have been reported during typical storage periods of these devices [283, 284]. The presence of crystals can pose several problems in the performance of transdermal systems. The release of the drug from the patch will be affected by the relative rates of dissolution of the crystals into the solution phase of the drug through the adhesion. [285]. Furthermore, the possibility of the existence of a modified chemical form of the drug leads to concerns about the effectiveness of the transdermal device in delivering the correct therapeutic agent.

Variankaval et al. [286] used a variety of characterization tools to determine the physical and chemical nature of the precipitates formed in situ in estradiol patches. Optical microscopy revealed that crystals were formed in a single layer inside the adhesive matrix and that they were present in two different morphologies: needle-like crystals and aggregates around the needles. From IR measurements, it was evident that estradiol probably was present in more than one crystal form in these patches. Raman microscopy showed that the needle-like crystals contain the adhesive component and the aggregates some modified crystal form of estradiol, indicating that, in addition to the drug, the polymeric adhesives also crystallize during storage [286].

Thacharodi and Rao reported permeation-controlled transdermal drug delivery systems using chitosan [287–290]. Studies on propranolol hydrochloride (prop-HCl) delivery systems using various chitosan membranes with different cross-link densities as drug release-controlling membranes and chitosan gel as the drug reservoir were performed. The physicochemical properties of the membranes to both lipophilic and hydrophilic drugs have been reported [287, 289]. In vitro evaluations of the TDS devices while supported on rabbit pinna skin were carried out in modified Franz diffusion cells [289]. The in vitro drug release profiles showed that all devices released prop-HCl in a reliable and reproducible manner. The drug release was significantly reduced when cross-linked chitosan membranes were employed to regulate drug release in the devices. Moreover, the drug release rate was found to depend on the cross-link density within the membranes. The device constructed with a chitosan membrane with a high cross-link density released a minimum amount of drug. This is due to the decreased permeability coefficient of the cross-linked membranes resulting from the cross-link points. Apart from the mentioned studies, the following points should be taken into consideration when fabricating a TDS. The major limitations of the TDS route are the difficulty of permeation of the drug through human skin and skin irritation [282, 291]. Studies have been carried out to find safe permeation enhancers to improve the transdermal flux of drugs [292]. Non-ionic surfactants, especially polysorbates, have been found to be safe permeation enhancers for the transdermal permeation enhancement of a variety of drugs [293].

Rajendran et al. [294] investigated a modified matrix for the in vitro delivery of terbutaline sulfate through excised guinea pig skin, using non-ionic surfactants as permeation enhancers. The flux of terbutaline sulfate from transdermal patches containing any of the selected non-ionic surfactants or without surfactants was determined using a Keshary-Chein cell. Among the spans used, span 80 produced the highest permeation of the drug. Adequate levels of transdermal permeation were observed with these agents [294].

7
Polymer Tablets

Tablets are still considered as the dosage forms of choice for reasons such as low manufacturing cost and good stability. Many direct compression diluents have been reported, but every diluent has some disadvantages [295]. Crystalline cellulose (MCC) has been widely used as a tablet diluent in Japan. Chitin and chitosan, on account of their versatility, were reported to be useful diluents for pharmaceutical preparations [296–298].

7.1
Pectin and Hydroxypropyl Methylcellulose

Oral cavity membranes are potential sites of drug administration, as there is excellent access and the drug can be applied, localized, and removed easily. Reports are available in which a variety of drugs have been shown to be absorbed by the mucosal epithelium of the oral cavity, mainly by the buccal or sublingual mucosa [299]. Miyazaki et al. [300] designed and evaluated both single and bilayer tablets of pectin and HPMC for the sublingual delivery of diltiazem. A significant improvement of the bioavailability of diltiazem when administered sublingually to rabbits from tablets (with pectin:HPMC = 1:1) was observed. Furthermore, the plasma concentration-time curves for sublingual tablets showed evidence of sustained-release of drug (5 h vs. 1 h) and also satisfactory hardness with good bioadhesion to the rat peritoneal membrane [300]. Also, in vitro studies of bilayer tablets showed improved controlled rates [300].

7.2
Chitosan Tablets

Sawayanagi et al. [301] reported the fluidity and compressibility of combined powders, viz., lactose/chitin, lactose/chitosan, potato starch/chitin, potato starch/chitosan as well as the disintegration properties of tablets made from these powders in comparison with those of combined powders of lactose/MCC and potato starch/MCC. This was done in order to develop direct compression dilutents as a part of their studies on pharmaceutical applications of chitin and chitosan.

From their investigations, it appears that the fluidity of combined powders with chitin and chitosan was greater that that of the powder with crystalline cellulose. The reported hardness of the tablets follows the order chitosan tablets > MCC > chitin. In the disintegration studies, tablets containing less than 70 % chitin or chitosan have passed the test. Moreover, the ejection force of the tablets of lactose/chitin and lactose/chitosan was significantly smaller than that of lactose/MCC tablets [301]. However, no reports are available on CDR formulations using these studies.

Recently, chitosan is gaining importance as a disintegration agent due to its ability to absorb water well. Reports reveal that chitosan contained in tablets at levels below 70 % acts as a disintegration agent [302]. Nigalayae et al. [303] investigated the sustained release characteristics of chitosan in the presence of citric acid or carbomer-934P in tablets containing theophylline as the model drug [303]. The rate of the drug release was slower in the tablets containing citric acid or carbomer-934P, used as anionic complex agents, than in those containing chitosan alone.

Mi et al. [304] reported chitosan tablets for the controlled release of theophylline. Alginate was used as the anionic polyelectrolyte to control the swelling and erosion rate of the chitosan tablets in acidic medium. Investigations on drug release mechanism of various tablets were carried out following Peppas's model [305, 306] and a nuclear magnetic resonance imaging microscopy was also introduced to examine the swelling/diffusion mechanism of various tablets [304].

7.3
Ethylcellulose Matrix Tablets

EC is an inert, hydrophobic polymer that has been studied substantially for its application as a matrix-forming material in direct compression tablets. Direct compression is the preferred method of manufacture for producing tablets intended for immediate or sustained release. There have been reports on the compressibility and compatibility of EC [307–310] and on its use as a matrix-forming material in direct compression tablets for the delivery of soluble and poorly soluble drugs [311–316]. Tablet hardness [310, 314, 315], the particle size of the polymer [314, 315, 317], and the viscosity grade [313, 314] were observed to directly affect the drug release rate. It was noted that tablet hardness affected the dissolution half-life more profoundly than did the viscosity grade [313]. Lower viscosity grades were more compressible and allowed a wider range of tablet hardness, and thus dissolution rates, for theophylline and indomethacin [313]. It was found that there is apparently a limit to the tablet hardness that could be obtained by an increase in compression force when a particular viscosity grade is employed [314, 315]. One of the major problems associated with hydrophobic matrix tablets is the reduction in the terminal release rate. The erosion of the EC matrix over time can serve to lessen this problem [316].

The mechanism of the drug release appeared to be simple diffusion from an EC matrix tablet and the data could be adequately described by the Higuchi square root of time relationship for water-soluble pseudophedrine hydrochloride at 12.5–25 % drug loading [309]. Release of the slightly soluble theophylline or the practically insoluble indomethacin from such tablets at 50 % or 25 % drug loading, respectively, was described by diffusion with polymer relaxation and erosion contributions to the release mechanism [313].

Recently, investigations were carried out to analyze the release data from EC matrix tablets to determine which release equation provides the best fit to the data and to observe the effect of drug solubility on the release mechanism [318]. Tablets were prepared by direct compression of drug, EC, and lubricant in an appropriate mass ratio to achieve a high and a low drug loading. Theophylline, caffeine, and dyphylline were selected as non-electrolyte xanthine derivatives with solubilities from 8.3 to 330 mg/ml at 25 °C. Drug release studies were performed at 37 °C in water with UV detection at 272 nm. Several equations to characterize release mechanisms were tested with respect to the release data. Drug diffusion, polymer relaxation, and tablet erosion were the mechanisms considered. The Akaike Information Criterion was also considered to ascertain the best-fit equation. At high drug loading the drug was released by a diffusion mechanism with a rate constant that increased with an increase in aqueous stability. At low drug loading, polymer relaxation also becomes a component of the release mechanism. However, its contribution to drug release was less pronounced as the solubility decreased, becoming negligible in the case of theophylline [318]. At each drug loading, an increase in drug solubility resulted in an increase in the dissolution rate, but did not change the best-fit release model [318].

7.4
Hydroxypropyl Cellulose Tablets

Cellulose ethers are hydrophilic polymers that are widely used as a pharmaceutical excipient [319]. They are generally considered to be stable in the solid state when kept in closed containers under normal environmental conditions [320] and the standards established by official pharmacopoeias for their storage are accordingly not particularly stringent: The United States Pharmacopoeia [321] and British Pharmacopoeia [322] merely require that they can be stored in closed containers, without specifying further measures for prevention of water uptake or microbiological contamination. The microorganisms most frequently isolated from pharmaceutical raw materials are those with very frugal nutrient requirements and good tolerance to dryness: notably *Bacillus, Streptococcus,* Gram-negative bacteria, yeasts, and molds [323]. The risks they pose for the integrity of cellulose ethers have hitherto received little attention [324, 325], even though cellulose ethers with a moderate degree of substitution are often used in culture media for cellulose-producing fungi and bacteria because they constitute a suitable source of carbon for these microorganisms [326–329]. The importance of the risk of excipient contamination derives from likelihood that it will significantly alter the properties of the dosage forms in which the excipient is subsequently incorporated [323, 330, 331].

Alvarez-Lorenzo et al. [332] studied the stability of several varieties of HPC during long-term storage, concentrating on the properties that are most important for their use as pharmaceutical excipients [333–336] and investigated the effects of any changes in these properties on drug release from direct compression tablets formulated with the selected HPLC. Theophylline was used as a model drug in these studies. The stability of several varieties of HPC was monitored during 3 years of storage (1) under the conditions recommended by manufacturer and Official pharmacopoeias (simple storage in closed containers), (2) at zero relative humidity. After 1 year, severe degradation of the varieties with lower initial pH and particle size stored at ambient relative humidity was shown by changes in their molecular weight and in pH and apparent viscosity of 2 % aqueous dispersions. Microbiological analyses showed the observed degradation to be attributable to the action of fungi of the genus *Rhizomucor*. The changes in apparent viscosity significantly affected the release of theophylline from direct compression tablets formulated with degraded excipient [332].

8
Miscellaneous

8.1
Mucoadhesive Chitosan Coated Liposomes

Mucoadhesive dosage forms have been received substantial attention as novel delivery systems to improve the bioavailability of drugs by prolonging the residence time and controlling drug release characteristics. A possible approach to obtain a specific bioadhesive drug delivery system could be the use of chitosan coated liposomes [337]. Recently, lecithin liposomes with/without FITC dextran, were prepared by the ethanol injection method and coated with three different molecular weight chitosans [337]. All the chitosan-coated liposomes were of spherical shape and no morphological differences between coated and uncoated liposomes were observed. The highest entrapment was found for liposomes coated with medium molecular weight chitosan. The stability of chitosan-coated liposomes in stimulated gastric fluid was significantly higher as compared to uncoated liposomes, which reveals the importance of chitosan as a stabilizing agent [337].

8.2
Chitosan-Lipid Emulsions

Positively charged systems are receiving more attention as novel colloidal drug carriers for various potential therapeutic applications [338]. Groth et al. [339] and Cortesi et al. [340] reported a novel use of cationic liposomes to target DNA into the cell nucleus and this allows the replacement of the defective gene product and thus restores the normal cell function. Furthermore, it is well known from emulsion and liposomal studies that the surface charge and the size of the colloidal carrier may effect the biofate of a drug in various organs of the body following i.v. administration [341, 342]. Moreover, in many papers Benita and his coworkers [343, 344] showed the possibility of producing stable, positively charged submicron emulsions, which are assumed to display several advantages. Davis et al. [345] suggested that positively charged emulsion droplets can behave differently, when introduced into the bloodstream, to normal (negatively charged) fat emulsion droplets with respect to the uptake of plasma blood components and opsonic factors. Therefore, positively charged systems may alter the pharmacokinetic profile of the incorporated drugs, resulting in a possible drug targeting with enhanced local drug concentration in the organs [345].

It would also be interesting to explore the intrinsic effects of the positively charged emulsions in accessible organs such as skin or cornea, which are known to carry a net negative charge [346, 347]. So far all positively charged, submicronized emulsions were based on a mixture of phospholipids with Poloxamer and stearylamine, as a cationic emulsifier. Unfortunately, stearylamine showed a high toxicity against the tested cell system in vitro [348–350]. This cytolytic and cytotoxic activity limits the utilization of the advantages of these systems as novel drug delivery carriers [351].

Jumma and Muller [352] reported the physicochemical properties of chitosan-lipid emulsions and their stability during the autoclaving process. The intention of this investigation was to formulate a stable, positively charged emulsion with a non-toxic cationic polymer (chitosan) resulting in an improved emulsion system for drug delivery. The experiments were carried out and optimized using an experimental design in order to estimate the appropriate concentrations of chitosan and the non-ionic surfactant F68 (ABA block copolymer). A mixed film consisting of the ABA block copolymer and chitosan molecules was formed at the o/w interface with an overall positive surface charge. Conversely, a combination between chitosan and phospholipids and/or with a mixture of phospholipids with the ABA block copolymer showed separation during autoclaving. A chitosan type with a low viscosity was used which was intended for a possible use in ocular and parenteral applications. An experimental factorial design 3(2) was used to investigate the effect of chitosan and F68 concentrations on the physicochemical properties of the system and consequently their influence on the stability of emulsions during autoclaving. Both size and surface charge of emulsions were significantly affected as a function of the chitosan concentration. They could achieve formulation with a mean particle size ranging from 125 to 130 nm and with a positive surface charge of 20–23 mV. Moreover, the chitosan emulsions were autoclaved without a significant change in their particle size. It was concluded from their investigations that the increasing concentration of chitosan needs a higher amount of F68 in order to achieve stable emulsions during autoclaving, due to the interaction between the positively charged chitosan and the negatively charged, free fatty acids, which are contained in the oil phase (castor oil) [352].

8.3
Poly(DL-lactide-co-glycolide)-methoxypoly(ethylene Glycol) Copolymers

The application of block copolymers of biodegradable PLA and PLGA with PEG in controlled drug delivery and drug targeting has been proposed [352–355]. The triblock copolymers PLA-PEG-PLA and PLGA-PEG-PLGA form more hydrophilic matrices than PLA and PLGA and are considered to be more suitable for controlled delivery of proteins [356]. Also, diblock PLA-mPEG and PLGA-mPEG copolymers have been used in the preparation of nanoparticles, exhibiting prolonged residence in the systemic circulation after intravenous administration and therefore having a potential as targeted drug carriers [357]. These polymers are not yet commercially available. Although the synthesis of PLA-PEG-PLA triblock copolymers has been studied [356, 358–361] and a coordinative reaction mechanism has been proposed for the polymerizations catalyzed by stannous octoate [362], relatively few reports appear to exist on the synthesis of PLGA-mPEG diblock copolymers [363]. Moreover, none of these studies deals in a systematic way with the effect of the conditions of preparation on the properties of the prepared PLGA-mPEG-copolymers.

Belesti et al. [364] reported a systematic study of the effect of certain preparative variables, such as the composition of the feed and the polymerization on the properties of the resulting PLGA-mPEG copolymers. The results with regard to the molecular weight and yield were discussed in relation to the polymerization mechanism proposed by Du et al. [362]. It was found that the higher the PEG contents of the feed the lower were the molecular weight of the copolymer and the yield of the reaction. The breadth of the molecular weight distribution decreased initially with time, but appeared to stabilize later at low values. Both the ethylene oxide content and the lactide to glycolide molar ratio in the copolymer appear to depend on the reaction temperature and varied with reaction time. From these investigations, PLGA and mPEG are partially miscible, and copolymers containing approximately 40 mol % or more ethylene oxide exhibit crystallinity [364].

8.4
Polymer-Based Gene Delivery Systems

Recently, gene therapy has gained a significant interest due to its ability to treat human disease by correcting the genetic deficiency of key metabolic enzymes

[365]. Despite the high transfection efficiency of viral vectors, their successful uses have been hampered due to immunogenicity, potential infectivity, complicated production, and inflammation [366]. Synthetic non-viral vectors are being widely sought as potential alternatives and cationic liposomes have been extensively used in vitro and in vivo [367]. The use of PDMAEMA was reported as an efficient non-viral transfectant [368, 369]. This polymer is able to bind electrostatically to plasmids yielding polymer-plasmid complexes. It was found that the size of the complexes formed was a dominant factor for the transfection efficiency. The highest transfection efficiency was observed at a polymer/plasmid ratio of 3 (w/w) and average molecular weight of the polymer above 250 KDa. Under these conditions particles with a size between 0.15 and 0.25 µm and a slightly positive γ-potential were formed. However, the complexes have a limited stability in aqueous solution due to possible chemical and physical degradation processes. In their further studies, to preserve the size and transfection potential of polymer-plasmid complexes, Cherng et al. [370] used freeze-drying and freeze-thawing for stabilization of the complexes. It was found that the concentration of the sugars is an important factor affecting both the size and transfection capability of the complexes after freeze-drying and freeze-thawing. However, the type of lycoprotectant (sugar) used is of minor importance. In these studies, it was also shown that when damage to polymer-plasmid complexes occurs, it results from the drying process but is not due to the freezing step [370].

8.5
Polypeptides Containing γ-Benzylglutamic Acid

Synthetic polypeptides continue to be of interest as drug delivery platforms because of their capacity to be both biocompatible as well as biodegradable to naturally-occurring biological products [371]. The large number of amino acids along with their range of physical and chemical properties renders this class of polymers useful for the design of novel drug delivery systems. By appropriate selection of amino acid residues and sequences, a variety of polypeptides can be designed to possess varying degrees of hydrophobicity, structural attributes, and electrostatic properties. This feature offers the potential that specific polypeptide polymers possessing the requisite physicochemical properties could ultimately be tailor-made to yield optimum delivery for particular drugs. Two obstacles are commonly cited as limiting the suitability of polypeptides for drug delivery applications: their biological degradation and potential toxicity [371, 372]. A number of approaches has been reported addressing these potential pitfalls. Because of the stability of the amide backbone towards hydrolysis, the catalytic activity of native enzymes is usually relied upon for polypeptide degradation [373, 374]. An alternative approach is to incorporate labile chemical bonds into the polypeptide backbone, thereby introducing hydrolytic instability to the polymer chain. To this end, the synthesis of copolymers of amino acids with either α-hydroxy acids [375] or anhydrides [376] has been described. The concerns over polypeptide toxicity are generally minimized by careful selection of amino acid residues and by limiting the number of different amino acids in the polypeptide backbone [371, 377]. The enhanced stability of polypeptides can influence the mechanisms and kinetics of drug release by affecting the time scales over which both drug diffusion and polymer degradation occur [378]. Drug release from hydrolytically labile polymers such as PLGA can proceed by both a diffusion controlled mechanism [379] and a degradation controlled mechanism [380–382]. Thus, variations in polymer degradation rates could lead to unreproducible drug release patterns thereby affecting both the efficacy and toxicity profile of the drug therapy. In addition, the onset of polymer degradation could cause structural deterioration of the implanted device. As a result, the risk of dose-dumping is increased should either the partially degraded system collapse or require retrieval when unexpected reactions occur during treatment. For these reasons, uncoupling the mechanism of drug release from polymer degradation would be a desirable alternative. Polypeptide polymers, which are less prone to hydrolysis and can, thus, provide structural integrity over a long period of time during which drug release may proceed through a diffusion-controlled mechanism, appears to be an outstanding choice.

Yang and co-workers have done a considerable amount of work on polypeptides as drug delivery platforms. One of their earlier findings demonstrated that simple changes in the ratio of hydrophilic to hydrophobic amino acid residues within the polypeptide backbone could modify the diffusion-controlled drug release properties of such a material [383].

Yang and workers reported the synthesis of three modified polypeptides containing γ-benzylglutamic acid as the common structural backbone. PBLG was chosen as the model polypeptide in their studies because of its helical structure in the solid state [384]. The structural attributes of this polymer were modified by random copolymerization of the D- and L-isomers of γ-benzylglutamic acid to produce PBDLG. In addition, the bulk polymer hydrophobicity of PBLG was modified by polymerization of PBLG blocks (A) with a hydrophilic pre-polymer block of PEG (B) to from an ABA triblock copolymer (PBLG-PEG-PBLG). The physicochemical properties of these three synthetic polypeptides were characterized and drug release studies were performed using two model hydrophilic drugs, viz., procainamide hydrochloride and protamine sulfate [383]. PDLG displayed a significantly slower release of procianamide when compared to PBLG, whereas the ABA triblock copolymer exhibited much faster release rates for both procianamide and protamine than those demonstrated by the other two polymers. The results indicate that, by using ABA, protamine release rates ranging from 2 weeks to approximately 2 months were obtained by simply the varying the polymer processing conditions and protein particle size. A nearly complete release of protein was obtained from ABA and this occurred without reliance upon degradation of the polymer backbone [39].

8.6
Aliphatic Polyanhydrides

Aliphatic polyanhydrides derived from fatty acids have been used as carriers for controlling drug delivery [385–387]. The common aliphatic diacids in medically used polyanhydrides were SA and FAD, with little attention being paid to other diacid monomers of this series. The elimination of biodegradable polymer implants from the body involves the degradation of the polymer into water-soluble degradation products, which are carried away and excreted. Polyanhydrides are composed, in general, of sparingly water-soluble diacid monomers and thus elimination via dissolution may be a slow process. The elimination of copolyanhydrides of sebacic acid with P(CPP-SA) or with P(FAD-SA) have been thoroughly investigated [385–388]. P(CPP-SA) wafers implanted in the brain of rats and rabbits showed that the relatively water soluble comonomer, SA (solubility in water, 1 mg ml^{-1}) was eliminated within 1 week leaving the less soluble comonomer CPP (0.01 mg ml^{-1}) to be excreted over 4–6 weeks [388]. Similarly, SA was released from P(FAD-SA) implanted in bone within 2 weeks, while the FAD component was eliminated over a three-month period [389]. These studies indicate that solubilization of the degradation products in body fluids is a key parameter for the elimination of a biodegradable implant; however, a systematic study on this issue has not been reported. The effect of hydrolytic enzymes in the degradation rate of implantable polymers is essential for understanding the in vivo behavior of an implanted polymer matrix. The degradation of PLA and poly(glycolide) in the presence of several hydrolytic enzymes found around the implants has been studied [390, 391]. Under the conditions employed, bromelain and enzymes with esterase activity significantly influenced the polymer degradation. The effect of enzymes on the degradation of polyanhydrides was not studied.

Apart from these, Albertsson and Lundmark [392] prepared PAA, PSA, and PDA from the mixed anhydrides of diacids and ketene in THF, the mixed anhydrides being subjected to melt polycondensation. The polymerizations were carried out in the presence of diethylzinc as a catalyst. The polymer yields and molecular weights of the polymers increased in the presence of catalyst and a mechanism for the reaction was proposed on the basis of the results obtained. The highest melting temperature of the resultant polymers was obtained from PDA (Tpeak=90 °C). The heat of fusion was approximately 100 J/g (PDA). The number-average and weight-average molecular weights of the polyanhydrides were 18000 and 33000, respectively (PDA). PA-β-PEG and PA-β-PTMG copolymers were prepared from the mixed anhydrides of dodecanoic acid, PEG-succinic acid half-ester and PTMG-succinic half-ester, respectively. The hydrolytic degradation of PDA has been studied in a buffered salt solution (pH 7.2) at 37 °C by examining the changes in molecular weight, heat of fusion, mass loss, and surface structure of melt-pressed PDA strips.

Domb and Neudelman [393] studied biopolyanhydrides using a polymer series of linear aliphatic diacids, with the following objectives: to study systematically the effect of monomer solubility in the degradation and in vivo elimination of polyanhydrides; (2) to evaluate the use of linear aliphatic natural diacids as components in drug delivery implants with increasing degradation rates; and (3) to determine the effect of common body enzymes on the degradation process of polyanhydrides. The polymer series are expected to degrade into their monomer counterparts at about the same rate but differ in the water solubility of their degradation products. Polymers based on natural diacids of the general structure $-[OOC-(CH_2)_x-CO]_n-$, where x is between 4 and 12 were implanted subcutaneously in rats and the elimination of the polymers from the implant site was observed. The in vitro hydrolysis of this polymer series was studied by monitoring the weight loss, released monomer degradation products, and the content of anhydride bonds in the polymer as a function of time. Dependence was found between the monomer solubility and the rate of polymer elimination both in vitro and in vivo. It was observed that the elimination time for polymers based on soluble monomers (x = 4 – 8) was 7–14 days, while the polymers based on low monomer solubility (x = 10 – 12) were eliminated only after 8 weeks. The in vitro degradation of polyanhydrides in the presence of several common hydrolytic enzymes found around implants did not affect polymer degradation. They found these polymers to be biocompatible and useful carriers for drug delivery [393].

In a similar fashion, Teomim and Domb [394] reported a systematic study on the synthesis, characterization, degradation, and drug release of fatty acid terminated PSA. A second class of fatty acid-based polyanhydrides synthesized from non-linear hydrophobic fatty acids based on MA, SA, and ricinoleics possessed the desired physicochemical properties such as melting point, hydrophobicity, and flexibility in addition to biocompatibility and biodegradability. The polymers were synthesized by melt condensation to yield film-forming polymers with molecular weights exceeding 50,000. The drug release rate from these fatty acid-containing polymers was significantly slowed and a constant drug release for months was achieved. Similarly, in their procedure, fatty acid-terminated sebacic acid polymers were synthesized by melt condensation of acetate anhydrides of linear fatty acids (C_8-C_{18}) and sebacic anhydride oligomers to yield waxy, off-white materials. Polymers with molecular weights (Mw) in the range of 9000 to 5000 were obtained for 10 % and 30 % (weight ratio) containing fatty acid terminals, respectively. Up to about 30 % of fatty acid terminals, the final product is mainly fatty acid-terminated polymer with up to about 5 % w/w of the symmetrical fatty anhydride. It is observed that increasing amounts of fatty acid acetic acid anhydride in the polymerization mixture had little effect on the polymer molecular weight up to a ratio of 40 : 60 (FAA:SA oligomer), the MW remaining in the range of 5000–8000. Above this ratio the molecular weight dropped to a level of 2000–3000 and the percent of the symmetrical anhydride increased to 10–40 %. Detailed investigations demonstrated that the fatty acid terminals had little effect on PSA melting and crystallinity. However, the fatty acid terminals had a slight effect on polymer degradation and drug release rate. PSA with 30 % w/w of $C_{14}-C_{18}$ terminals degraded and released the incorporated drug for more that 4 weeks as compared with 10 days for the acetate-terminated PSA [394].

8.7
Ricinoleic Acid-Based Biopolymers

It is clear from Sect. 8.6 that the delivery of drugs from polyanhydrides has been extensively studied. Fatty acids are good candidates for the preparation of biodegradable polymers, as they are natural body components [395], and their hydrophobicity allows retention of the encapsulated drug for longer time periods when used as drug carriers. However, the monofunctionality of fatty acids restricted their use to serve as monomers for polymerization. As discussed in the previous section, fatty acids are converted to monomers by dimerization of unsaturated fatty acids such as oleic acid and erucic acid. Their homopolymers are viscous liquids; copolymerization with increasing amounts of SA forms solid polymers with increasing melting points (30–70 °C) as a function of SA content. The in vitro and in vivo drug release characteristics, toxicity, and elimination of these polymers have been reported [396–399]. Poly(FAD-SA), prepared by mixing the drug in the melted polymer, has shown promising results in treating brain tumors in laboratory animals [400]. The same polymer has been used for the delivery of gentamicin sulfate for the treatment of osteomyelitis [401, 402]. Gentamicin was released for a few weeks both in vitro and in vivo. Continuous glutamic acid stimulation of trigeminal motoneurons, using poly(FAD-SA) microspheres, showed a pronounced effect on the developing skeleton of growing rates [267].

Although these polymers were found to be suitable for drug delivery applications, in vivo studies in dogs showed that, when implanted in muscle, the polymer degraded to semisynthetic FAD monomers, which slowly cleared (6 months) from the implantation site [386]. This erucic acid-based FAD is not easily metabolized in vivo, probably due to the C-C linkage between the two fatty acids.

Castor oil is a mixture of triglycerides, predominantly of ricinoleic acid (85–90 %), which has an unusual structure with a double bond at the 9-position and a hydroxy group in the 12-position (*cis*-12-hydroxyoctadeca-9-enoic acid) [403]. It is one of the few commercially available glycerides that contain a hydroxy functionality in such a high percentage on one fatty acid. Reports are available on copolyesters of citric acid and castor oil synthesized by standard polycondensation under vacuum at an elevated temperature using anhydrous $FeCl_3$ as the catalyst [404, 405]. Drug release studies were conducted using sulfadiazine [404] and paracetamol [405] as model drugs and zero order release kinetics were observed.

Teomim et al. [406] reported the synthesis of ricinoleic acid and hydrogenated ricinoleic acid (hydroxystearic acid)-based monomers, which were synthesized from the attachment of a carboxylic acid side chain via a hydrolyzable ester bond to the hydroxy group (succinic acid/maleic acid). These monomers were copolymerized with sebacic acid and tested for their suitability as drug carriers [406]. In vitro studies showed that these polymers underwent rapid hydrolytic degradation in 10 days. Moreover, methotrexate release from the polymers was not affected by the initial polymer molecular weight in the range of 10,000–35,000. From these investigations it appears that the in vitro drug release correlated with the degradation of the polymers, and the fatty acid ester monomers were further degraded to their counter parts, ricinoleic acid and succinic/maleic acid.

8.8
Water-Soluble Polyamides

Neuse and coworkers [407–410] have extensively studied water-soluble polyamides as potential drug carriers. Some interesting features of their studies include the grafting of PEO chains onto drug carrier type polysapartamides [407], with a view to increase the carrier's hydrophobicity, reduce immunogenicity, and enhance resistance to both protein binding and capture by RES with concomitant prolongation of the residence time in serum circulation. Such features have proved to be of outstanding biomedical benefit for drug carrier molecules [411, 412]. In addition to the PEO grafts, the polymers contained amine-functionalized side groups for drug conjugation. With these amine functions linked to the polymer backbone by very short (less than 10 atomic constituents) spacers, any attached drug species would reside in close proximity to the main chain and remain embedded in the protective layer of PEO "tentacles" surrounding the backbone. While this spatial arrangement offers certain pharmacokinetic advantages, it also creates difficulties associated with poor accessibility of the biofissionable polymer-drug to proteolytic enzymes instrumental in the intracellular (lysosomal) release of the drug from the carrier.

8.9
Poly(trimethyl Carbonate)-poly(adipic Anhydride) Blends

Blending provides a neat and smooth means of combining desirable properties of different polymers. Biodegradable matrices with new combinations of polymer properties and modification of drug release profiles can thus be obtained [413]. The use of degradable polymers has been favored because they eliminate the need for surgical removal after depletion. Linear aliphatic polycarbonates, such as PMTC, have shown to be suitable for these applications, being biocompatible and degradable by simple hydrolysis, promoted in vivo by enzymatic activity [414]. PTMC displays high elasticity at room temperature but degrades slowly in aqueous solution, showing little molecular weight loss, sample weight loss, or change in morphology after several months [415]. Attempts have been made to change the chemical composition by copolymerization of TMC with ε-caprolactone or D,L-lactide or by blending PTMC with other degradable homopolymers [416–419]. Edlund and Albertsson [420] described the copolymerization in bulk and solution of TMC with AA as well as the blending of homopolymers. Drug delivery from the blends was evaluated. They proposed a statistical factorial model to explore the influence of three important blend parameters and their interactions, making it possible to predict the erosion and drug release behavior of the blend matrices. It was found that the PAA:PTMC ratio and molecular weight of the polycarbonate component significantly influence the drug-release performance, mass loss, and degree of plasticization. The interaction among these factors is expected to influence the blend properties. Therefore, from these studies it is evident that blending offers a convenient alternative to copolymerization for the preparation of polymer matrices with predictable drug delivery [420].

9
Conclusion

This article has summarized in detail the status of the field of biopolymers and drug delivery. The progress in this field is characterized by a steady accumulation and build-up of concepts and information throughout the past three decades that has helped to develop new drug delivery products for clinical use. Researchers in the field had to adopt and find solutions for the emerging biotechnology, therapeutic agents, sensitive proteins, oligo- and polynucleotides, and cells. Compared to the efforts invested in studying and developing drug delivery systems, the clinical outcome is disappointing. Only a few invasive polymeric systems have reached the clinic, using mainly one type of polymer, poly-lactide-polyglycolide. Drug delivery systems in the clinic include injectable polylactide-based microspheres releasing LHRH analogues, somatostatin, and recently growth hormone for periods of one month or longer. Another device is the BCNU releasing implant for treating brain tumors and an injectable PLA solution for treating head and neck cancer and periodontal disease. The main use, however, for drug delivery systems so far has been in oral extended release systems that have already become a standard for many drugs. This review has described dozens of degradable and non-degradable polymers and delivery formulations ranging from nanospheres, microparticles, films and tablets that might and should be developed into clinical products. Some of the systems and concepts are at early stage of practical development with a great potential to become a viable technology if proper investment and dedication is made. A major part of this survey is dedicated to nano-size and micro-size particulate systems. These particulate systems have been prepared from a long list of polymers and natural components with different properties. The incentive for developing and studying these particulate systems has been their need and potential use as carriers of drugs, cells, and genes. This article provides a background for the new emerging field of nanotechnology that will have an impact on the development of sophisticated nano-size delivery systems. The field is now ripe for technology development moving from the bench to the clinic. Still, there is a great need for breakthroughs in this multidisciplinary field of research and technology for better target active agents, to improve the delivery of gene therapy agents and sensitive proteins. This will emerge from collaborative teamwork involving chemists pharmacists, pharmacologists, toxicologists, physicians, and other relevant scientists.

References

1. Domb AJ, Benitolila A (1998) Acta Polym 49:526
2. Hayashi T (1994) Prog Polym Sci 19:663
3. Hsieh D, Rhine W, Langer R (1983) J Pharm Sci 72:12
4. Yashida R, Saka K, Okano T, Sakurai Y (1991) Jpn J Artif Organs 2:465
5. Dong DC, Hoffman AS (1990) Proc Int Symp Controlled Release Bioact Mater 17:325
6. Ishihara K, Hamada N, Kato S, Shinohara I (1984) J Polym Sci Polym Chem Ed 22:881
7. Suzuki A, Tanaka T (1990) Nature 346:345
8. Kwon IC, Bae YH, Kim SW (1991) Nature 54:291
9. Kokufata E, Zhang YG, Janaka J (1991) Nature 351:302
10. Kost J (1990) Pulsed and self drug delivery. CRC Press, Boca Raton
11. Diepold R, Kreuter J, Guggenbuhl P, Robinson JR (1989) Int J Pharm 54:149
12. Illum L, Wright J, Davis SS (1989) Int J Pharm 52:221
13. Douglas SJ, Davis SS, Illum L (1997) CRC Crit Rev 3:233
14. Kreuter J (1991) J Controlled Release 16:169
15. Kramer PA (1974) J Pharm Sci 63:1647
16. Roser M, Kissel T (1993) Eur J Pharmacol Biopharm 39:8
17. Bhargava K, Aindo HY (1992) Pharm Res 9:776
18. Rajaonaryvony MJ, Vauthier C, Couarraze G, Puisieux F, Couvreur P (1993) J Pharm Sci 82:912
19. Couvreur P, Vatuthier C (1991) J Controlled Release 17:187
20. Allemann E, Gurny R, Doelker E (1993) Eur J Pharm Biopharm 39:173
21. Gref R, Minamitake Y, Peracchia MT, Trubetskoy V, Torchlin V, Langer R (1994) Science 263:1600
22. Seijo B, Fattal E, Roblot-Treupel L, Couvreur P (1990) Int J Pharm 62:1
23. Cifcti K, Suheyla Kas H, Atilla Hincal A, Meral Ercan T, Guven O, Ruacan S (1996) Int J Pharm 131:73
24. Witschi C, Doelker E (1997) Eur J Pharm Biopharm 4:215
25. Jeon HJ, Jeong Y, Jang MK, Park YH, Nah YW (2000) Int J Pharm 207:99
26. Yoo HS, Lee KH, Oh JE, Park TG (2000) J Controlled Release 68:419
27. Santos-Magalhaes NS, Pontes A, Pereira VMW, Caetano MNP (2000) Int J Pharm 208:71
28. Jagur-Grodzinski J (1999) React Funct Polym 39:99
29. Kwon GS, Okano T (1996) Adv Drug Del Rev 21:107
30. Scholz C, Iijima M, Nagasaki Y, Kataoka K (1995) Macromolecules 26:7295
31. La SB, Okano T, Kataoka K (1996) J Pharm Sci 85:85
32. Zhang X, Burt HM (1995) Pharm Res 12:S265
33. Dunn SE, Coombes AGA, Garnett MC, Davis SS, Davis MC, Illum LJ (1997) J Controlled Release 44:65
34. Moghimi SM, Muir IS, Illum L, Davis SS, KolBacnofen V (1993) Biochim Biophys Acta 1179:157
35. Allock HR, Ducher SR, Scopelianos AG (1994) Biomaterials 15:563
36. Allock HR, Fuller TJ, Mack DP, Matsumura K, Smeltz KM (1997) Macromolecules 10:824
37. Vandorpe J, Schacht E, (1997) Biomaterials 18:1147
38. Gref R, Domb AJ (1995) Adv Drug Del Rev 16:215

39. Hrkach JS, Peracchia MT, Domb AJ, Lotan N, Langer R (1997) Biomaterials 18:27
40. Coombes AGA, Tasker S, Lindblad M, Holmgreen K, Hoste V (1997) Biomaterials 18:1153
41. Stella B, Arpicco S, Peracchia MT, Desmacle D, Hoebeke J, Renoir M, D'Angela J, Cattel L, Couvreur P (2000) J Pharm Sci 89:1452
42. Damage C, Michel C, Aprahamian M, Couvreur P, Devissaguet JP (1990) J Controlled Release 13:233
43. Damage C, Hillaire-Buys D, Puech R, Hoeltzel A, Michel C, Ribes G (1995) Diab Nutr Metab 8:3
44. Alkhouri Fallouh N, Roblot Treupel L, Fessi H (1986) Int J Pharm 28:125
45. Leonard F, Kulkarni RK, Brandes G, Nelson J, Cameron J (1996) J Appl Polym Sci 10:259
46. Damge C, Vranckx H, Balschmidt P, Couvreur P (1997) J Pharm Sci 86:1403
47. Aboubakar M, Puisieux F, Couvreur P, Vauthier C (1999) Int J Pharm 183:63
48. Lambert G, Fattal E, Pinto-Alphandary H, Gulik A, Couvreur P (2000) Pharm Res 17:707
49. Kwon GS, Suwa S, Yokoyana M, Okano T, Sakuri Y, Kataoka K (1994) J Controlled Release 29:17
50. Oh I, Lee K, Kwon H-Y, Lee Y-B, Shin S-C, Cho C-S, Kim C-K (1999) Int J Pharm 181:107
51. Cho C-S, Kim S-W, Komoto T (1997) Macromol Rapid Commun 191:981
52. Jeong YI, Cheon JB, Kim S-H, Nah JW, Lee Y-M (1998) J Controlled Release 51:169
53. Kreuter J (1998) Biomed Sci Tech p.31
54. Amiji M, Park K (1994) In: Shalaby SW, Ikada Y, Langer R, Williams J (eds), Polymers of biological Significance. ACS Symposium Series 540, Washington, DC
55. Lehr CM, Boustra JA, Schacht EH, Junginer HE (1992) Int J Pharm 78:43
56. Calvo P, Remunan-Lopez C, Vila-Jato JL, Alonso MJ (1997) J Appl Polym Sci 63:125
57. Chang J, Zhongguo SY (1996) Chinese J Biomed Eng 15:102
58. Kaplun AP, Son LB, Krasnopolsky YM, Shvets VI (1999) Vopr Med Khim 45:3
59. Schwarz C, Mehnert W (1999) J Microencapsulation 16:205 and references cited therein
60. Kim YS, Kim KS (1998) Yakehe Hakhoechi (Korean) 28:249
61. Mitra R, Pezron I, Chu WA, Mitra AK (2000) Int J Pharm 205:127
62. Miguel ID, Imbertie L, Rieumajou V, Major M, Kravtzoff R, Betbeder D (2000) Pharm Res 17:817
63. Marengo E, Cavalli R, Caputo O, Rodriguez L, Gasco MR (2000) Int J Pharm 205:3
64. Ogata N (1998) Polym Prep 170:225
65. Ravi Kumar MNV, Kumar N (2001) Drug Dev Ind Pharm 27:1
66. Chen LH (1998) Pharm Dev Technol 3:241
67. Hoffman AS, Afracsiake A, Dong LC (1996) J Controlled Release 4:212
68. Siegel RA, Falamarzian M, Firestone BA, Moxley BC (1988) J Controlled Release 8:179
69. Bronsted J, Kopecek J (1991) Biomaterials 12:584
70. DeMoor CP, Doh L, Sigel RA (1991) Biomaterials 12:836
71. Apocella A, Cappello B, Del Nobile MA et al. (1993) Biomaterials 14:83
72. Harris JM (1992) In: Harris JM (ed), Poly(ethylene glycol) chemistry. Biotechnology and biomedical applications. Plenum Press, New York, Chapter 1
73. Katre NV (1993) Adv Drug Delivery Rev 10:91
74. Graham NB (1992) In: Harris JM (ed), Poly(ethylene glycol) chemistry. Biotechnology and biomedical applications. Plenum Press, New York, pp 263–281
75. Embry MP, Graham NB (1990) US Patent 4,894,238
76. Graham NB, McNeill ME (1984) Biomaterials 5:27
77. Bromberg L (1996) J Appl Polym Sci 59:459
78. Langer R (1990) Science 249:1527
79. Kulkarni RK, Pani KC, Neuman C, Leonard F (1966) Arch Surg 93:839
80. Domb AJ, Gallardo CF, Langer R (1989) Macromolecules 22:3200
81. Heller J, Sparer RV, Zentner GM (1991) In: Chaisin M, Langer R (eds), Biodegradable polymers as drug delivery systems. Dekker, New York, p 171

82. Negishi N, Bennett DB (1987) Pharm Res 4:305
83. Couvreur P, Puisieux F (1993) Adv Drug Delivery Rev 10:142
84. Spenlehauer G, Vert M, Benoit JP, Chabot F, Veillard M (1988) J Controlled Release 7:217
85. Steber W, Fishbein R, Cady SM (1987) Eur Patent Appl 87,111,217
86. Bennett DB, Adams NW, Li X, Kim SW (1988) J Bioact Compatible Polym 3:44
87. Wakiyama N, Juni K, Nakano M (1982) Chem Pharm Bull 30:3719
88. Cho CS (1996) J Appl Polym Sci 60:161
89. Pitt CG, Gratzei MM, Kimmei GL, Surles J, Schindler A (1981) 2:215
90. Ramaraj B, Radhakrishnan G (1994) J Appl Polym Sci 52:837
91. Abe K, Koide M, Tsuchida E (1977) Macromolecules 10:1259
92. Wu PC, Huang YB, Fang JY, Tsai YH (1998) Drug Dev Ind Pharm 24:179
93. Jvazquez M, Perez-Marcos B (1992) Drug Dev Ind Pharm 18:1355
94. Sung KC, Nixon PR (1996) Int J Pharm 142:53
95. Gao P, Nixon PR, Skoug JW (1995) Pharm Res 12:965
96. Lu G, Jun HW (1998) Int J Pharm 84:344
97. Alvarez-Lorenzo C, Gomez-Amoza JL, Martinez-Pacheco R, Souto C, Concheiro A (1999) Int J Pharm 180:91
98. Schmolka IR (1972) J Biomed Mater Res 6:571
99. Miller SC, Donovan MD (1982) Int J Pharm 12:147
100. Miyazaki S, Takeuchi S, Yokouchi C, Takada M (1984) Chem Pharm Bull 32:4205
101. Miyazaki S, Yokouchi C, Nakamura T, Hashiguchi N, Hou WM, Takada M (1986) Chem Pharm Bull 34:1801
102. Choi HG, Jung JH, Ryu JM, Yoon SJ, Oh YK, Kim CK (1998) Int J Pharm 165:33
103. Jain NK, Shah BK, Taneja LN (1991) Indian J Pharm Sci 53:16
104. Morikawa K, Okada F, Hosokawa M, Kobayashi H (187) Cancer Res 47:37
105. Miyazaki S, Ohkawa Y, Takada M, Attwood D (1992) Chem Pharm Bull 40:2224
106. Attwood D, Collett JH, Davies MC, Tait CJ (185) J Pharm Pharmacol 37:5P
107. Altinok H, Yu GE, Nixon K, Gorry PA, Attwood D, Booth C (1997) Langmuir 3:5837
108. Tanodekaew S, Deng NJ, Smith S, Yang YW, Attwood D, Booth C (1993) J Phys Chem 97:11847
109. Yuguchi Y, Mimura M, Urakawa H (1997) Proc Int Workshop on Green Polymers – Reevaluation of Natural Polymers, Indonesia, pp 306–329
110. Miyazaki S, Suisha F, Kawasaki N (1998) J Controlled Release 56:75
111. Suisha F, Kawasaki N, Miyazaki S (1998) Int J Pharm 172:27
112. Kawasaki N, Ohkura R, Miyazaki S (1999) Int J Pharm 181:227
113. Bywater S (1979) Adv Polym Sci 30:90
114. Kanaoka S, Sawamoto M, Higashimura T (1991) Macromolecules 24:2309
115. Kanaoka S, Sawamoto M, Higashimura T (1992) Macromolecules 25:6407
116. Sogah DY, Hertler WR, Webster OW, Cohen GM (1987) Macromolecules 20:1473
117. Bazan CC, Schrock RR (1991) Macromolecules 24:817
118. Ruckenstein E, Zhang H (1999) Macromolecules 32:3979
119. Buchholz FL (1994) Chem Br (Aug) 652
120. Chen J, Park H, Park K (1999) J Biomed Mater Res 44:53
121. Muzzarelli RAA, Baldassarre V, Conti F. et al. (1988) Biomaterials 9:247
122. Hou WM, Miyazaki S, Takada M, Komai T (1985) Chem Pharm Bull 33:3986
123. Ravi Kumar MNV (2000) React Funct Polym 27:1
124. Michaels AS, Miekka RG (1961) J Phys Chem 65:1765
125. Sakiyama T, Chu CH, Fujii T, Yano T (1993) J Appl Polym Sci 50:2021
126. Yao KD, Peng T, Xu MX (1994) Polym Int 34:213
127. Yao KD, Peng T, Goosen MFA (1993) J Appl Polym Sci 48:343
128. Yao KD, Peng T, Feng HB, He YY (1994) J Polym Sci Part A Polym Chem 32:1213
129. Peng T, Yao KD, Goosen MFA (1994) J Polym Sci Part A Polym Chem 32:519
130. Yao KD, Lin J, Zhao RZ, Wang WH, Wei L (1998) Die Angewandte Makromol Chemie 255:71

131. Kim SS, Lee YM, Cho SS (1995) J Polym Sci Part A Polym Chem 33:2285
132. Kim SS, Lee YM, Cho CS (1996) 36th IUPAC Int Symp on Macromol. Seoul, Korea,
133. Lee YM, Kim SS, Kim SH (1997) J Materi Sci: Mater Med 8:537
134. Dutta PK, Viswanathan P, Mimrot L, Ravi Kumar MNV (1997) J Polym Mater 14:537
135. Ravi Kumar MNV, Singh P, Dutta PK (1999) Indian Drugs 36:393
136. Ravi Kumar MNV, Singh P, Dutta PK (1999) Eastern Pharm June : 109
137. Ravi Kumar MNV, Dutta PK, Nakamura S (2000) Indian J Pharm Sci (Jan) 55
138. Dutta PK, Ravi Kumar MNV (1999) Indian J Chem Technol 6:55
139. Zhang X, Macdonald DA, Goosen MFA, Macauley (1994) J Polym Sci Chem Ed 32:2965
140. Zhu KJ, Xianghou L, Shilin Y (1990) J Appl Polym Sci 1:39
141. Lee YM, Kim SS (1997) Polymer 38:2415
142. Kim SY, Lee YM, Lee SI (1997) Int Symp Controlled Release Bioact Mater. Stockholm, Sweden
143. Cho SM, Kim SY, Lee YM, Sung YK, Cho CS (1999) J Appl Polym Sci 73:2151
144. Yao KD, Yin YJ, Xu MX, Wang YF (1995) Polym Int 38:77
145. Qu X, Wirsen A, Albertsson AC (1998) ACS Polymeric Materials Science and Engineering 79:242
146. Wood JM, Attwood D, Collett JH (1982) J Pharm Pharmacol 34:1
147. Wood JM, Attwood D, Collett JH (1981) Int J Pharm 7:189
148. Wood JM, Attwood D, Collett JH (1984) In: Hoffman AS, Ratner BD, Horbett TA (eds), Polymers as biomaterials. Plenum Press, New York, pp 347–360
149. Lasic DD (1990) J Colloid Interface Sci 140:302
150. Florence AT (1993) In: Gregoriadis G (ed), Liposome technology, CRC Press, Boca Raton, FL, Vol 2, pp 157–176
151. Uchegbu IF, Vyas SP (1998) Int J Pharm 172:33
152. Yoshioka T, Florence AT (1994) Int J Pharm 108:117
153. Murdau S, Gregoriadis G, Florence AT (1996) STP Pharm Sci 6:44
154. Yoshioka T, Strenberg B, Florence AT (1994) Int J Pharm 105:1
155. Reddy DN, Udupa N (1993) Drug Dev Ind Pharm 19:843
156. Uchegbu IF, Bouwstra JA, Florence AT (1992) J Phys Chem 96:10548
157. Uchegbu IF, Double JA, Kelland LR, Turton JA, Florence AT (1996) J Drug Targeting 3:399
158. Uchegbu IF, McCarthy D, Schatzlein A, Florence AT (1996) STP Pharm Sci 6:33
159. Uchegbu IF, Florence AT (1995) Adv Colloid Interface Sci 58:1
160. Uchegbu IF, Schatzlein A, Vanlerberghe G, Morgatini N, Florence AT (1997) J Pharm Pharmacol 49:606
161. Bernard MS, Arunothayanum P, Uchegbu IF, Florence AT (1996) 3rd UKCRS Symposium on Controlled Drug Delivery, Manchester, UK
162. Ono A, Yamaguchi M, Horikoshi L, Shintani T, Uneo M (1994) Biol Pharm Bull 17:166
163. Nishimura K, Nishimura S, Seo H, Nishi N, Tokura S, Azuma I (1986) J Biomed Mater Res 20:1359
164. Kinemura Y (1987) Fine Chem 12:5
165. Ida G, Monica C, Annalia A, Bice C, Luisa M (1998) Carbohydr Polym 36:81
166. Hayashi T, Ikada Y (1991) J Appl Polym Sci 42:85
167. Chandy T, Sharma CP (1992) Biomaterials 13:949
168. Chandy T, Sharma CP (1993) Biomaterials 14:939
169. Chandy T, Sharma CP, Sunny MC (1987) J Biomat Appln 1:533
170. Yao KD, Peng T, Yu JJ, Xu MX, Goosen MFA (1995) JMS Rev Macromol Chem Phys C35:155
171. Mathiowitz E, Saltzman WM, Domb A, Dor P, Langer R (1988) J Appl Polym Sci 35:755
172. Mathiowitz E, Langer R (1987) J Controlled Release 5:13
173. Brown LR, Wei CL, Langer R (1983) J Pharm Sci 72:1181
174. Wheatley MA, Chang M, Park E, Langer R (1991) J Appl Polym Sci 43:2123
175. Murata Y, Toniwa S, Miyamoto E, Kawashima S (1999) Int J Pharm 176:265
176. Dunn EJ, Zhang X, Sun D, Goosen MFA (1993) J Appl Polym Sci 50:353

177. Huguet ML, Groboillot A, Neufeld RJ, Poncelet D, Dellacherie E (1994) J Appl Polym Sci 51: 1427
178. Huguet ML, Dellacherie E (1996) Process Biochemistry 31:745
179. Huguet ML, Neufeld RJ, Dellacherie E (1996) Process Biochemistry 31:347
180. Maurice DM (1989) In: Saetlone MS, Bucci G, Speiser P (eds), Fida research series. Liviana Press, Padova, vol 2
181. Seig JW, Robinson JR (1975) J Pharm Sci 64:931
182. Sanzgivi YD, Maschi S et al. (1993) J Controlled Release 26:195
183. Albertsson AC, Carlfors J, Sturesson C (1996) J Appl Polym Sci 62:695
184. Kang KS, Veeder GT, Mirrasoul PJ, Kaneko T, Cottrell IW (1982) Appl Environ Microbiol 43:1086
185. Doner LW, Douds BD, (1995) Carbohydr Res 273:225
186. Jansson PE, Lindberg B, Sandford PA (1983) Carbohydr Res 124:135
187. O'Neill MA, Selvendran RR, Morris NJ (1983) Carbohydr Res 124:123
188. Kuo MS, Mort AJ, Dell A (1986) Carbohydr Res 156:173
189. Chandrasekaran R, Radha A, Thailembal VG (1992) Carbohydr Res 224:1–17
190. Sanderson GR, Clark RC (1983) Food Technol 37:62
191. Milas M, Shi X, Rinando M (1990) Biopolymers 30:451
192. Grasdalen H, Smidsrod O (1987) Carbohydr Polym 7:371
193. Deasy PB, Quigley J (1991) Int J Pharm 73:117
194. Anderson DMW, Brydon WG, Eastwood MA (1988) Food Addit Contam 5:237
195. Shungu D, Valiant M, Tutlane V, Weinberg E (1983) Appl Environ Microbiol 46:840
196. Harris JE (1985) Appl Environ Microbiol 50:1107
197. Colegrove GT (1983) Ind Eng Chem Prod Res Dev 22:456
198. Rozier A, Mazuel C, Grove J, Plazonnet B (1997) Int J Pharm 153:191
199. Sanzgiri YD, Maschi S, Crescenzi V, Callegaro L (1993) J Controlled Release 26:195
200. Chilvers GR, Morris VJ (1987) Carbohydr Polym 7:111
201. Alhaique F, Carafa M, Coviello T, Murtas E (1995) Proc Controlled Release Soc 22:286
202. Alhaique F, Carafa M, Coviello T, Murtas E, (1996) J Controlled Release 42:1981
203. Santucci E, Alhaique F, Carafa M, Coviello T (1996) J Controlled Release 42:157
204. Kedzierewicz F, Lombry C, Rios R, Hoffman A, Maincent P (1999) Int J Pharm 178:129
205. Daumas-Duport C, Tucker ML, Kolles JH, Cervera P, Beuvon F, Varlet P (1997) J Neuro Oncol 34:61
206. Djordjevic B, Snuybalski K (1960) J Exp Med 112:509
207. Menei P, Benoit JP, Biosdron-celle M, Fournier D, Mercies P, Guy G (1994) Neurosurgery 34:1058
208. Benoit JP, Marchais H, Rolland H, Van de Velde V (1996) In: Benita S (ed), Microencapsulation methods and industrial applications. Marcel Dekker, New York, pp 35–72
209. Bodmeir R, Chen H, Davidson RGW, Hardee GE (1997) Pharm Dev Technol 2:323
210. Shenderova A, Burke TG, Schwendeman SP (1997) Pharm Res 14:1406
211. Gubskaya AV, Lihnyak YV, Blagoy YP (1995) Drug Dev Ind Pharm 21:1953
212. Annapragada A, Adjei A (1996) Int J Pharm 136:1
213. Villiers VM, Tiedt LR (1996) Pharmazie 51:564
214. Geze A, Verier-Julienne MC (1999) Int J Pharm 178:257
215. Rojas J, Pinto-Alphandary H (1999) Int J Pharm 183:67
216. Blanco-Prieto MJ, Bewsseghir K (1999) Int J Pharm 184:243
217. Ricordi C, Tzakis AG et al. (1992) Transplantation 53:407
218. Warnock GL, Knetemoan NM, Ryan EA, Rabinovitch A, Rajotte ARV (1992) Diabetoligia 35:89
219. Pyzdrowski KL, Kendall DM et al. (1992) N Engl J Med 327:220
220. Chang TMS (1964) Science 146:524
221. Limand F, Sun AM (1980) Science 210:908
222. Weber CJ, Zabinski S (1990) Transplantation 49:396
223. Wijsman J, Atkinson P (1992) Transplantation 54:588

224. Chicheportiche D, Reach G (1988) Diabetologia 31:54
225. Schrezenmeir J, Gero L et al. (1992) Transplant Proc 24:2925
226. Lum ZP, Krestow M, Tai IT, Vacek I, Sun AM (1992) Transplantation 53:1180
227. Halle JP, Leblond FA (1994) Cell Transplant 3:365
228. Pariseau JF, Leblond FA (1995) J Biomed Mater Res 29:1331
229. Robitaille R, Pariseau JF (1999) J Biomed Mater Res 44:116
230. Ohya Y, Takei T (1993) Chem Ind (Jpn) 46:798
231. Yin Y, Xu M, Chen X, Yao KD (1996) Chinese Sci Bull 41:1266
232. Gupta KC, Ravi Kumar MNV (1999) JMS Pure Appl Chem A36:827
233. Gupta KC, Ravi Kumar MNV (2000) Polym Int 49:141
234. Gupta KC, Ravi Kumar MNV (2000) Biomaterials 21:1115
235. Gupta KC, Ravi Kumar MNV (2001) J Appl Poly Sci 80:639
236. Gupta KC, Ravi Kumar MNV (2000) J Appl Polym Sci 76:672
237. Gilding DK (1981) In: Williams DF (ed), Biocompatibility of clinical implant materials, CRC Press Inc, Boca Raton, FL, Vol II, p 8
238. Doddi N, Versfeldt CC, Wasserman D (1977) US Patent 4,052,988
239. Casey DJ, Roby MS (1984) Eur Patent EP 098394 AI
240. Albertsson AC, Sjoling MJ (1992) J Macromol Sci Chem A29:43
241. Mathisen T, Masus K Albertsson AC (1989) Macromolecules 22:3842
242. Albertsson AC, Lofgren A (1992) Makromol Chem Macromol Symp 53:221
243. Albertsson AC, Lundmark S (1990) J Macromol Sci Chem A27:397
244. Palmgren R, Karlsson S, Albertsson AC (1997) J Polym Sci A Polym Chem 35:1635
245. Albertsson AC, Lofgren A, Sturesson C, Sjoling M (1994) In: Ottenbrite RM (ed), Polymeric Drugs, Drug Administration. ACS Symposium Series 545, Chapter 14
246. Wakiyama N, Juni K, Nakano M, (1982) Chem Pharm Bull 30:3719
247. Haynes D, Kirkpatrick A (1985) Anesthesiol 63:490
248. Langerman L, Golomb E, Benita S (1991) Anesthesiol 74:105
249. Mashimo T, Uchida I, Park M, Shibata A, Nishimura S, Inagaki Y, Yoshiya I (1992) Anesth Analg 74:827
250. Masters DB, Domb AJ (1998) Pharm Res 15:1038
251. Domb AJ (1993) US Patent 5,188,837
252. Cousins MJ, Bridenbaugh PO (1988) In: Lippincott JB (Ed), Neural blockade in clinical, anesthesia, and management of pain, Philadelphia, pp 121–128
253. Domb AJ, Maniar M (1991) In: AAPS Meeting, Washington DC, 1991
254. Amselem S, Alving CR, Domb AJ (1992) Polym Adv Technol 3:351
255. Amsellem S, Domb AJ, Alving CR (1992) Vaccine Res 1:383
256. McGeer PL, Eccler JC, McGeer EG (1987) Molecular neurobiology of the mammalian brain. 2nd edn, Plenum, New York
257. Chandler SH (1989) Brain Res 477:252
258. Byrd KE, Stein ST, Sokoloff AJ, Shankar K (1990) Am J Anat 189:93
259. Byrd KE, Domb AJ, Sokoloff AJ, Hamilton-Byrd EL (1992) Polym Adv Tech 3:337
260. Byrd KE, Hamilton-Byrd EL (1994) In: Domb AJ (ed), Polymeric site-specific pharmacotherapy. John Wiley & Sons, New York, pp 142–155
261. Howard MA(III), Gross A, Grady MS, Langer RS (1989) J Neurosurg 71:105
262. Kubek MJ, Rea MA, Hodes ZI, Aprison MH (1983) J Neurochem 40:1307
263. Palkorits M, Mezey E, Eskay RL, Brownstein MJ (1986) Brain Res 373:246
264. Merchenthaler I, Csernus V, Csontos C, Petrusz P, Mess B (1988) Am J Anat 181:359
265. Sharif NA (1989) Ann NY Acad Sci 553:147
266. Rekling JC (1990) Brain Res 510:175
267. Byrd KE, Sukay MJ, Dieterle MW, Yang L, Marting TC, Teomim D, Domb AJ (1997) J Dent Res 76:1437
268. Takahashi T, Takayama K, Machida Y, Nagai T (1990) Int J Pharm 61:35
269. Michaels AS (1965) Ind Eng Chem 57:32
270. Bodmeier R, Wang J (1993) J Pharm Sci 82:191

271. Benita S, Benoit JP, Puisieux F, Thies C (1984) J Pharm Sci 73:1721
272. Lee KY, Park WH (1997) J Appl Polym Sci 63:425
273. Kratz F, Fichtner I, Beyer U, Schumacher P, Roth T, Fiebig HH, Unger C (1997) Eur J Cancer 33:S175
274. Matsumura Y, Maeda H (1986) Cancer Res 46:6387
275. Takakura Y, Fujita T, Hashida M, Sezaki H (1990) Pharm Res 7:339
276. Katti D, Krishnamurti N (1999) J Microencapsulation 16:231
277. Nagai T (1985) J Controlled Release 2:121
278. Azuma J, Seto Y, Mochizuki N, Hamaguchi T, Nozaki Y (1991) Rinsyo-Iyaku 7:1921
279. Deck DE (1997) Eur J Surg 63:49
280. Park H, Robinson JR (1985) J Controlled Release 2:47
281. Matsuda S, Iwata H, Se N, Ikada Y (1999) J Biomed Mater Res 45:20
282. Chien YW (1981) Transdermal controlled systemic medication. Marcel Dekker, New York, vol 31
283. Stefano F, Biali F, Scasso A (1997) Proc Int Symp Controlled Release Bioact Mater 24:703
284. Carrig T, Therriault S (1997) Proc Int Symp Controlled Release Bioact Mater 24:891
285. Kubota K, Twizell EH, Maibach HL (1994) J Pharm Sci 83:1593
286. Variankaval NE, Jacob KI, Dinh SM (1999) J Biomed Mater Res 44:397
287. Thacharodi D, Panduranga Rao K (1993) J Chem Technol Biotechnol 58:177
288. Thacharodi D, Panduranga Rao K (1993) Int J Pharm 96:33
289. Thacharodi D, Panduranga Rao K (1995) Biomaterials 16:145
290. Panduranga Rao K (1998) Pure Appl Chem 70:1283
291. Behel CR, Barret M, Flynn GL (1981) J Pharm Sci 71:229
292. Rojanasakul Y, Hsieh DS (eds) (1994) Drug permeation enhancement: theory and applications. vol 62. Marcel Dekker Inc, New York
293. Shahi V, Zatz JL (1978) J Pharm Sci 67:789
294. Rajendran D, Prabushankar GL, Dhanraj SA, Dubey R, Suresh B (1996) Indain J Pharm Sci Nov-Dec:251
295. Seki H, Kagami T, Hayashi T, Okusa N (1981) Chem Pharm Bull 29:3680
296. Sawayanagi Y, Nambu N, Nagai T (1979) 99th Annual Meeting of the Pharm Soc Jpn, Sapporo
297. Bruscato FN, Danti AG (1978) US Patent 4,086,335
298. Capozza R (1975) German Patent 2,505,305
299. Harris D, Robinson JR (1992) J Pharm Sci 81:1
300. Miyazaki S, Kawasaki N, Nakamura T, Iwatsu M, Hayashi T, How WM, Attwood D (2000) Int J Pharm 204:127
301. Sawayanagi Y, Nambu N, Nagai T (1982) Chem Pharm Bull 30:2935
302. Ritthidej GC, Chomto P, Pummangura S, Menasveta P (1994) Drug Dev Ind Pharm 20:2019
303. Nigalaye A, Adusumilli P, Bolton S (1990) Drug Dev Ind Pharm 16:449
304. Mi, FL, Her NL, Kaun CY, Wong TB, Shyu SS (1997) J Appl Polym Sci 66:2495
305. Korsmeyer RW, Peppas NA (1984) J Controlled Release 1:89
306. Peppas LB, Peppas NA (1989) J Controlled Release 8:267
307. Shlieout G, Zessin G (1994) Eur J Pharm Sci 2:187
308. Upadrashta SM, Katikaneni PR (1994) Int j Pharm 112:173
309. Katikaneni PR, Upadrashta SM, Neau SH, Mitra AK (1995) Int J Pharm 123:119
310. Shlieout G, Zessin G (1996) Drug Dev Ind Pharm 22:313
311. Shaikh NA, Abidi SE, Block LH (1987) Drug Dev Ind Pharm 13:1345
312. Shaikh NA, Abidi SE, Block LH (1987) Drug Dev Ind Pharm 13:2495
313. Upadrashta SM, Katikaneni PR, Hileman GA, Keshary PR (1993) Drug Dev Ind Pharm (1993) 19:449
314. Katikaneni PR, Upadrashta SM, Rowlings CE (1995) Int J Pharm 117:13
315. Pollock DK, Sheskey PJ (1996) Pharm Technol 20:120

316. Pather SI, Russell I, Syce JA, Neau SH (1998) Int J Pharm 164:1
317. Dabbagh MA, Ford JL, Rubinstein MH, Hogan JE (1996) Int J Pharm 140:85
318. Neau SH, Howard MA, Claudius JS, Howard DR (1999) Int J Pharm 179:17
319. Doelker E (1993) Adv Polym Sci 107:200
320. Handbook of Pharmaceutical Excipients (1994) 2nd edn, American Pharmaceutical Association – The Pharmaceutical Press, London
321. United States Pharmacopoeia 23 – The National Formulary 18 (1995) United States Pharmacopoeial Convention Inc, Rockville
322. British Pharmacopoeia (1998) Her Majesty's Stationary Office, London
323. Dela Rosa MC, Medina MR, Vivar C (1995) Pharm Acta Helv 70:227
324. Banker G, Peck G, Williams E, Taylor D, Pirakitikulr (1982) Drug Dev Ind Pharm 8:41
325. Kasulke U, Philipp B, Polter E (1987) Acta Biotechnol 7:147
326. Arai M, Sakamoto R, Murao S (1987) Agric Biol Chem 51:627
327. Shambe T, Ejembi O (1987) Enzyme Microb Technol 9:308
328. Maskova HP, Vasilyeva LV, Kofronova O, Kunc F (1988) Folia Microbial 33:482
329. El-Naghy MA, El-Katany MS, Attia AA (1991) Int Biodeterioration 27:75
330. Blair TC, Buckton G, Bloomfield F (1991) Int J Pharm 71:111
331. Beveridge EG (1992) In: Hugo WB, Russell AD (eds) Pharmaceutical Microbiology. Blackwell, Oxford
332. Alvarez-Lorenzo C, Duro R, Gomez-Amoza JL (1999) Int J Pharm 18:105
333. Dahl TC, Calderwood T, Bormeth A, Trumble K, Piepmeir E (1990) J Controlled Release 14:1
334. Acquier R, Belhani F, Maillos H, Delonca H (1992) STP Pharm Sci 2:469
335. Vazquez MJ, Perez-Marcos B, Gomez-Amoza JL et al. (1992) Drug Dev Ind Pharm 18:1355
336. Mitchell K, Ford JL, Amstrong DJ et al. (1993) Int J Pharm 100:143
337. Filipovic-Grcic F, Skalko-Basnet N, Jalsenjak I (2001) J Microencapsulation 18:3
338. Steger DL, Desnick JR (1977) Biochim Biophys Acta 464:530
339. Groth D, Keil O, Lehman C, Schneider M, Rudolp M, Reszka R (1998) Int J Pharm 162:143
340. Cortesi R, Esposito, Menegatti E, Gambari R, Nastruzzi C (1996) Int J Pharm 139:69
341. Juliano RL, Stamp D (1975) Biochim Biophys Commun 63:651
342. Laval-Jeantet AM, Laval-Jeantet M, Bergot C (1982) Intervent Radiol 17:617
343. Elbaz E, Zeevi A, Klang S, Benita S (1993) Int J Pharm 96:R1
344. Klang SH, Frucht-Pery J, Hoffman A, Benita S (1994) Int J Pharm 46:986
345. Davis SS, Illum L, Washington C, Harper G (1992) Int J Pharm 82:99
346. Klang SH, Baszkin A, Benita S (1994) Int J Pharm 108:57
347. Zeevi A, Klang S, Alard V, Brossard F, Benita S (1994) Int J Pharm 108:57
348. Adams DA, Richardson GJ, Ryman BE, Wisniewski HM (1977) J Neurol Sci 31:173
349. Layton D, Luckenbach GA, Andreesen R, Munder PG (1980) Eur J Cancer 16:1529
350. Cambell PI (1983) Cytobios 37:21
351. Filion MC, Phillips NC (1998) Int J Pharm 162:159
352. Jumma M, Muller BW (1999) Int J Pharm 183:175
353. Stolnik S, Dunn SE, Garnett MC (1994) Pharm Res 11:1800
354. Emile C, Bazile D, Herman F, Helene C, Veillard M (1996) Drug Delivery 3:187
355. Jeong B, Bae YH, Lee DS, Kim SW (1997) Nature:388–860
356. Youxin L, Kissel T (1993) J Controlled Release 27:247
357. Bazile D, Prud' Homme C, Bassoullet MT (1995) J Pharm Sci 84:493
358. Zhu K, Xianghou L, Shilin Y (1990) J Appl Polym Sci 39:1
359. Deng XM, Xiong CD, Cheng LM, Xu RP (1990) J Polym Sci Polym Lett 28:411
360. Jedlinski Z, Kurcok P, Walach W, Janeczek H, Radecka I (1993) Makromol Chem 194:1681
361. Kricheldrof HR, Meier-Haack J (1993) Makromol Chem 194:715
362. Du YJ, Lemstra PJ, Nijenhuis AJ, Van Aert HAM, Bastiaansen C (1995) Macromolecules 28:2124

363. Gopherich A, Gref R, Minanitake Y (1994) ACS Symposium Series 567, 242
364. Belesti A, Leontiadis L, Klepetsauis P, Ithakissios DS, Avgoustakis K (1999) Int J Pharm Sci 182:187
365. Mahato R, Rolland A, Tomlinson E (1997) Pharm Res 14:7
366. Smith A (1995) Ann Rev Microbiol 14:7
367. Park YK, Park YH, Shin BA, Choi ES, Park YR, Akaoke T, Cho CS (2000) J Controlled Release 69:97
368. Cherng JY, Wetering Pvd, Talsma H, Crommelin DJA, Hennink HE (1996) Pharm Res 13:1038
369. Wetering Pvd, Cherng JY, Talsma H, Crommelin DJA, Hennink WE (1998) J Controlled Release 53:145
370. Cherng JY, Wetering Pvd, Talsma H, Crommelin DJA, Hennink WE (1999) Int J Pharm 183:25
371. Andreson JM, Spilizewski KL, Hiltner A (1985) In: Williams DF (ed), Biocompatibility of Tissue Analogs. CRC Press, Boca Raton, FL
372. Sanders LM (1991) In: Lee VHL (ed), Peptide and protein drug delivery. Marcel Dekker, New York
373. Silman HI, Sela M (1967) In: Fasman (ed), Poly α-amino acids. Marcel Dekker, New York
374. Pytela J, Kotva R, Metalova M, Rypacek F (1990) Int J Biol Macromol 12:241
375. Katakai R, Goodman M (1982) Macromolecules 15:25
376. Domb AJ (1990) Biomaterials 11:686
377. Marck KW, Wildevur CRH, Sederel WL, Bantjes A, Feijen J (1977) J Biomed Mater Res 11:405
378. Hopfenberg HB (1982) In: Paul DR, Harris FW (eds), Controlled release polymeric formulations. ACS, Washington DC
379. Paul DR, McSpadden SK (1976) J Membr Sci 1:33
380. Lee PI (1980) J Membr Sci 7:255
381. Hutchinson FG, Furr BJA (1988) In: Johnson P, Lloyd-Jones JG (eds), Drug delivery systems. Ellis Horwood, Chichester
382. Shah SS, Cha Y, Pitt CG (1992) J Controlled Release 18:261
383. Cook T, Amidon GL, Yang VC (1997) Int J Pharm 159:197
384. Markland P, Amidon GL, Yang VC (1999) Int J Pharm 178:183
385. Domb AJ, Amselem S, Maniar M (1993) In: Langer R, Peppas NA (eds), Biopolymers. vol 107, Springer, Heidelberg, pp 93–141
386. Domb AJ, Amselem S, Langer R, Maniar M (1994) In: Shalaby S (ed), Designed to degrade biomedical polymers. Carl Hanser, p.69, USA
387. Domb AJ, Maniar M (1993) J Polym Sci 31:1275
388. Domb AJ, Rock M, Perkin C, Proxap B, Villemure JG (1994) Biomaterials 15:681
389. Shea J, Maniar M, Green M, Broxup B, Domb AJ, Rock M (1991) AAPS Meeting, Washigton DC
390. Williams DF (1981) Eng Med 10:5
391. Williams DF, Mort E (1977) J Bioeng 1:231
392. Albertsson AC, Lundmark S (1990) Br Polym J 23:205
393. Domb AJ, Neudelman R (1995) Biomaterials 16:319
394. Teomim D, Domb AJ (1999) J Polym Sci Part A Polym Chem 37:3337
395. Bremer J, Osmundensen H (1984) In: Numa S (ed), Fatty acid metabolism and its regulation. Elsevier, Amsterdam, pp 113–147
396. Brem H, Domb AJ, Lenartz D, Dureza C, Olivi A, Epstein JL (1992) J Controlled Release 19:325
397. Maniar M, Domb AJ, Haffer A, Shah J (1994) J Controlled Release 30:233
398. Tabada Y, Domb AJ, Langer R (1993) J Pharmacol Sci 83:5
399. Shieh L, Tamada, Tabada Y, Domb AJ, Langer R (1994) J Controlled Release 29:73
400. Olivi A, Ewend MG, Utsuki T, Tyler B, Domb AJ (1996) Cancer Chemother Pharmacol 39 (1996) 90

401. Domb AJ, Amselem S (1994) In: Domb AJ (ed), Polymeric site specific pharmacotherapy. Wiley, Chichester, pp 242–265
402. Laurencin C, Gerhart T, Witschger P (1993) J Orthop Res 11:256
403. Naughton FC (1993) In: Kroschitz JI, Howe-Grant M (eds), Encyclopedia of chemical technology. Wiley, New York, pp 301–320
404. Pramanick D, Biswas D, Ray TT, Bakr MDA (1994) J Polym Mater 11:41
405. Ravi Kumar MNV, Madhava Reddy, Dutta PK (1996) Iranian Polym J 5:60
406. Teomim D, Nyska A, Domb AJ (1999) J Biomed Mater Res 45:258
407. Neuse EW, Perlwitz AG, Barbosa AP (1994) J Appl Polym Sci 54:57
408. Machado M, Neuse EW, Perlwitz AG (1992) Angew Makromol Chem 35:35
409. Neuse EW (1994) Macromol Symp 80:111
410. Caldwell G, Neuse EW, Perlwitz AG (1997) J Appl Polym Sci 66:911
411. Herman S, Hooftman G, Schacht E (1995) J Bioact Compat Polym 10:145
412. Ohya Y, Kuroda H, Hirai K, Ouchi T (1995) J Bioact Compat Polym 10:51
413. Park TG, Cohen S, Langer R (1992) Macromolecules 25:116
414. Zhu KJ, Hendren RW, Jensen K, Pitt CG (1991) Macromolecules 24:1736
415. Albertsson AC, Eklund M (1995) J Appl Polym Sci 57:87
416. Albertsson AC, Eklund M (1994) J Polym Sci Part A Polym Chem 32:265
417. Buchholz B (1993) J Mater Sci Mater Med 4:381
418. Albertsson AC, Liu Y (1997) JMS Pure Appl Chem 34:1457
419. Edlund U, Albertsson AC (1998) Proc Int Symp Controlled Release Bioact Mater 25
420. Edlund U, Albertsson AC (1999) J Appl Polym Sci 72:227

Received: April 2001

Hydrogel-Based Colloidal Polymeric System for Protein and Drug Delivery: Physical and Chemical Characterization, Permeability Control and Applications

Ales Prokop[1], Evgenii Kozlov[1], Gianluca Carlesso[2], Jeffrey M. Davidson[2,3]

[1] 107 Olin Hall, Chemical Engineering Department, 24th Avenue South & Garland Avenue, Vanderbilt University, Nashville, TN 37235, USA
[2] Department of Pathology, Vanderbilt University School of Medicine, Nashville, TN 37232-2561, USA
[3] Research Service, Department of Veterans Affairs Medical Center, Nashville, TN 37212-2637, USA

The use of polymeric nanoparticles as drug carriers is receiving an increasing amount of attention both in academia and industry. The development of suitable delivery systems for protein drugs with high molecular weights and short half-lives is of current interest. In addition, nanoparticles have a number of potential applications in drug and vaccine delivery as well as gene therapy applications. This article features a new production technology for nanoparticles comprised of multicomponent polymeric complexes that are candidates for delivery vehicles of biological molecules such as proteins and drugs. Materials science theory and practice provide the basis for the development of highly compacted structures that are insoluble in water and buffered media. Biocompatible and mostly natural polymers are fabricated into thermodynamically stable nanoparticles, in the absence of organic solvents, using two types of processing: batch and continuous. Careful choice of construction materials and the superposition of several interacting principles during their production allow for the customization of the physicochemical properties of the structures. Among the typical polymers used to assemble nanoparticles, different polysaccharides, natural amines and polyamines were investigated. The entrapped substances tested included proteins, antigens and small drug molecules. The size and charge of nanoparticles is considered to be of primary importance for application in biological systems. Detailed experiments in batch and continuous systems allowed time-dependent stoichiometric characterization of the production process and an understanding of fundamental assembly principles of such supramolecular structures. Continuous-flow production is shown to provide more consistent data in terms of product quality and consistency, with further possibilities of process development and commercialization. To control permeability, polydextran aldehyde, incorporated into the particle core, was used to enable physiologic cross-linking and long-term retention of substances that would otherwise rapidly leak out of the nanoparticles. Results of cross-linking experiments clearly demonstrated that the release rate could be substantially reduced, depending on the degree of cross-linking. For vaccine antigen delivery tests we measured an antibody production following subcutaneous and oral administration. The data indicated that only the cross-linked antigen was immunogenic when the oral route of administration was used. The data presented in this paper address primarily the utility of nanoparticulates for oral delivery of vaccine antigen. This novel technology is extensively discussed in contrast to other technologies, primarily water- and organic solvent-based. The usefulness is demonstrated using several examples, evaluating protein and small drug delivery.

Keywords. Nanoparticle, Water-soluble polymers, Hydrophilic drugs, Protein delivery, Drug delivery, Antigen delivery, Processing, Permeability control

List of Abbreviations 121

1   Introduction 122

2   Materials and Methods 124

2.1   Materials. 124
2.2   Equipment and Procedures 126
2.2.1 Stoichiometry Evaluation in a Batch System 126
2.2.2 Stoichiometry Evaluation in a Continuous System. 128
2.2.3 Preparation of Nanoparticles Loaded with Model Proteins
       for Release Experiments. 128
2.2.4 Preparation of Nanoparticulate Vaccine System Loaded
       with Model Antigens. 129
2.2.5 Preparation of Nanoparticulate System Loaded
       with GMS-Polymer Conjugate 129
2.2.6 Aseptic Operation 131
2.2.7 Systems Used for Comparison 131
2.3   Encapsulation Efficiency and Drug Loading 131
2.4   Product Analysis (for Stoichiometry) 132
2.5   Measurement of the Primary Droplets. 132
2.6   Intermediate and Product Particle Characterization 132
2.7   Product Stability 133
2.8   Product Freeze-Drying 133
2.9   Release Measurement and Permeability Control 134
2.10   Compatibility and Animal Data 134
2.11   Immunizations and Immunoassays 135

3   Results 136

3.1   Rationale 136
3.2   Nanoparticle Assembly and Structure 136
3.3   Entrapment and Loading Efficiencies 137
3.4   Size and Charge Data 139
3.5   System Stoichiometry 145
3.6   Product Stability 146
3.7   Freeze-Drying of Product 148
3.8   Permeability Data 148
3.9   Vaccine Data 152
3.10   Biocompatibility Data 154

4   Discussion 154

4.1   General 154
4.1.1 Scope. 154
4.1.2 Uniqueness 155

4.1.3	Nanoparticulate Production Technologies (Based on Polyelectrolyte Complexes) 157
4.2	Special. 158
4.2.1	Novelty 158
4.2.2	Statistical Approach 159
4.2.3	Surface Properties and Steric Stabilization 160
4.2.4	Drug Loading and Entrapment Efficiency 161
4.2.5	Mechanism of Release: Matrix vs Reservoir-Type Device. 162
4.2.6	Mechanism of Release: Release from Hydrogels 162
4.2.7	Mechanism of Release: Release from Hydrogels Possessing Ionizable and Pendant Groups – Swelling 163
4.2.8	Mechanism of Release: Hydrophobic Interactions. 163
4.2.9	A Need for Protein Cross-Linking: Schiff Base 163
4.2.10	Release of Small Size Proteins and Molecules 164
4.2.11	Developing Safe Vaccine Vehicles: Bioavailability Issues 164
4.2.12	Comparison with Other Water-Based Technologies 166
4.2.13	Benefits of Controlled Delivery 168

References 169

List of Abbreviations

ACA	alkyl cyanoacrylate
BCA	bicinchoninic acid
BSA	bovine serum albumin
C	core polymer
CC	cytochrome C
CL	cross-linked
CRL-1005	adjuvant (CytRx Corp.)
CS	(sodium) cellulose sulfate
CSTR	continuous stirred tank reactor
CT	chitosan (glutamate) or chitosan sulfate
DL	drug loading, %
DNA	deoxyribonucleic acid
DT	diphtheria toxoid
EE	encapsulation efficiency, %
ELISA	enzyme-linked immunoassay
F38	Symperonic F38
F-68	polyethylene/polypropylene oxide copolymer
F68	Symperonic F68
bFGF	fibroblast growth factor, basic
GL	gellan
GMS	gentamycin sulfate
HMP	sodium hexametaphosphate
ID	internal diameter

IM	intramuscular
LDA	laser Doppler anemometry
MPHBL	Massachusetts Public Health Biologic Laboratory
MW	molecular weight
NIH	National Institutes of Health
NP, NPs	nanoparticle, nanoparticles
OR	oral
OVA	ovalbumin
PBS	phosphate-buffered solution
PCS	photon correlation spectroscopy
PDA	polydextran aldehyde
PE	polyelectrolyte (polymer)
PEC	polyelectrolyte complex
PEG	polyethylene glycol
PEO-PPO	polyethylene oxide-polypropylene
PMCG	poly(methylene-co-guanidine) hydrochloride
PVA	polyvinyl alcohol
QTI	Quantitative Technologies, Inc.
S	shell (corona) polymer
SA-HV	sodium alginate, high viscosity
SD	standard deviation
SP	spermine (tetrahydrochloride)
SQ	subcutaneous
TE	Tyndall effect
TPP	(pentasodium) tripolyphosphate
TT	tetanus toxoid
USP	US Pharmacopoeia
X-52	κ(kappa)-carrageenan

1
Introduction

New methods of drug administration can substantially reduce the development cost of new drug entities and they can result in an improved therapeutic index by modifying drug distribution and bioavailability. Reduced dosing frequency can improve patient compliance and minimize side effects. This can be accomplished by incorporating drugs into various delivery systems. Many macromolecules have a relatively short shelf life in aqueous medium, which can lead to both poor bioavailability and poor stability in the final product. The concept of polymeric nanoparticles (NPs) as drug carriers is receiving increasing attention both in academia and in industry. Nanoparticles have a wide variety of potential applications in the drug delivery, vaccine, and gene therapy areas. In particular, their delayed release kinetics can overcome the receptor down regulation that occurs with acute or burst doses of potential therapeutics such as epidermal growth hormone [1].

Various preparation techniques have been developed for nanoparticle production over the years. Choosing a preparation technique involves the consideration of several factors, the most important of which is the physicochemical properties of the polymer and the entrapped drug. The most common methods involve polymerization reactions, such as emulsion polymerization, dispersion polymerization and inverse microemulsion polymerization, using both degradable and non-degradable polymers, as reviewed by Vauthier-Holtzscherer et al. [2]. The value of these techniques is greatly diminished because of the use of an organic solvent or mineral oil. Even if the polymerization is accomplished in an acidic aqueous phase, as is case of cyanoacrylates [3], small residual molecular species (unreacted monomers, initiators and surfactants) may present a safety issue in the final product. Recently, poly(isobutyl cyanoacrylate) capsules, although biodegradable, have been shown to break down into potentially toxic products [4].

Other methods include use of preformed polymers, both natural and synthetic. Thermal denaturation in the case of protein-based nanoparticles [5], chemical cross-linking by glutaraldehyde and use of desolvating agents for gelatin-based nanoparticles [6] present potential hazards. In addition, methodologies within the same group include the use of organic solutions of polymers, solvent extraction [7] as well as evaporation [8], which all suffer from unfavorable chemistries and environments.

Development of new technologies to produce water-based drug delivery vehicles and strategies to optimize nanoparticle physiochemical properties, such as size, charge, and hydrophilicity, is receiving increasing attention. Methods incorporating hydrophilic compounds are few and difficult to implement. Among the most difficult tasks is the ability to prevent product aggregation. The challenges to formulate water-soluble drugs remain until the present time. Recent reviews summarize the state of the art [9–12].

This article discusses size and charge data for nanoparticles produced in both batch and continuous processes from two different polymer systems. In both systems, a polyanionic solution is atomized into a swirling bath of polycationic solution, forming nanoparticles as the solutions come together. This method of mixing solutions used to produce the nanoparticles is often referred to as "titration" (sequential addition of one polymer into another). Different ratios of polyanion to polycation were used to vary the composition of the nanoparticles, which were then evaluated for their size and charge.

Following a study of physical properties of these nanoparticles, further research focused on defining stoichiometric coefficients of reactants involved in a batch system. This study was undertaken to gain a better understanding of the assembly process for producing nanoparticles, mechanisms involved in drug release and control of their release. This process is then extensively compared with similar products on the market and reported in the literature.

2
Materials and Methods

2.1
Materials

Two basic systems were used in both the batch and continuous production processes for exploration of the main system parameters. They are referred to as Systems #1 to #2 throughout this article. System #1 contained pentasodium tripolyphosphate (TPP, anhydrous; Sigma, St. Louis, MO), κ(kappa)-carrageenan (X-52; Sanofi Bio-Industries, Waukesha, WI), chitosan glutamate (CT, Pronova Biopolymer, Drammen, Norway), calcium chloride (Sigma), Pluronic F-68 (Sigma), and ovalbumin (chicken egg albumin, OVA, grade V; Sigma) if the product is "loaded with protein". System #2 contained sodium alginate (SA-HV, high viscosity; Kelco, San Diego, CA), cellulose sulfate (CS; Janssen Chimica, Belgium), spermine tetrahydrochloride (SP, Sigma), poly(methylene-co-guanidine) hydrochloride (PMCG; Scientific Polymer Products, Inc., Ontario, NY), calcium chloride, Pluronic F-68, and ovalbumin if the product was "loaded". PMCG is a synthetic oligomer chemically similar to natural polyamines. F-68 is a polyethylene/polypropylene oxide copolymer and is used as a steric stabilizer [13]. The mass amounts of the individual components in solutions used to generate nanoparticles for Systems #1 and 2, divided into anionic and cationic (nonionic) compounds, are listed in Table 1. In this common case, denoted as standardized system, the anionic solution is referred to as a droplet-forming internal phase mixture (core polymer) and the cationic solution as a receiving bath mixture (corona or shell polymer). One example of the reverse chemistry is given in Table 2 (Example #8). This system is comprised of droplet-forming cationic and receiving bath anionic solutions.

Besides Systems #1 and #2, additional components were used to assemble other systems: gellan (GL; Sigma), heparin (from porcine intestinal musosa,

Table 1. The standardized chemistry used to make nanoparticles (g/100 mL)

System #1		System #2	
Anion (C)	Cation (S)	Anion (C)	Cation (S)
0.1 TPP	0.05 CT	0.05 HV	0.05 SP
0.1 X-52	0.1 Calcium chloride	0.05 CS	0.15 ml 50 % PMCG
0.4 Ovalbumin (if loaded)	1.0 Pluronic F-68 (non-charged)	0.5 Ovalbumin (if loaded)	0.05 Calcium chloride
–	–	–	1.0 Pluronic F-68 (non-charged)
100 mL water	100 mL water	100 mL water	100 mL water

C – core solution, S – shell (corona) solution.

Table 2. Examples of some other successful polymer chemistries

Example #	Core polymer solution	Corona (shell) polymer solution
1	HV, gellan, TT	PMCG, $CaCl_2$, F-68
2	HV, CS, 20 kbase plasmid (DNA)	PMCG, $CaCl_2$, F-68
3	HV, chondroitin sulfate C	PMCG, SP, $CaCl_2$, F-68
4	HV, chondroitic sulfate C, OVA	SP, PLL, $CaCl_2$, F-68
5	HV, CS, TT	PMCG, $CaCl_2$, F-68
6	Acacia gum, HV, OVA	BSA, $CaCl_2$
7	Kappa-carrageenan, OVA	CT, $CaCl_2$, F-68
8	Chitosan, PVA, $CaCl_2$, cationic protein	Heparin, F-68

Table 3. Average molecular weight of polymeric components

Component	Molecular weight (g mol^{-1})
Acacia gum (gum arabic)	250,000
Bovine serum albumin (BSA)	55,000
Carrageenan kappa (X-52)	N/A
Cellulose sulfate (CS)	1,200,000
Chitosan glutamate (CT)	128,000
Chondroitin sulfate	N/A
Cytochrome C	12,300
F-68	8,400
Fibroblast growth factor, basic (bFGF)	15,000
Gellan (GL)	N/A
Heparin	7,000
High viscosity sodium alginate (HV)	460,000
Low-ethoxylated pectin	N/A
Ovalbumin (OVA)	45,000
Pentasodium tripolyphospahe hexahydrate (TPP)	476
Poly(methylene-co-guanidine) hydrochloride (PMCG)	5,000
Polydextran aldehyde (PDA)	40,000
Polyvinylamine, hydrochloride (PVA)	15,000
Spermine hydrochloride (SP)	348
Tetanus toxoid (TT)	150,000

7 kDa, Sigma), tetanus toxoid (TT, lot #TAS319R8; Connaught Labs., Swiftwater, PA), cytochrome C (CC, Sigma), polyvinylamine (PVA, Air Products and Chemicals, Allentown, PA), polydextran aldehyde (PDA, Pierce, Rockford, IL), fibroblast growth factor-2 (bFGF, Scios, Inc., Mountain View, CA), sodium hex-

ametaphosphate (HMP, Fluka, Milwaukee, MI), low-ethoxylated pectin (Unipectine, 315 NH ND, Sanofi Bio-Industries, Paris, France) and high MW nonionic block copolymer CRL-1005 (11.5–12.5 kDa, CytRx Corp., Norcross, GA). The molar masses of components as obtained from the manufacturers are listed in Table 3.

Since we also present detailed results from other authors, some specifics are listed here. Polymers used by Calvo et al. [14,15] are listed below: chitosan (CT; Seacure® 123, viscosity 14 cps, Pronova Biopolymer, Drammen, Norway), polyethylene glycol (PEO, MW 4 and 10 k, Sigma), polyethylene oxide-polypropylene oxide block copolymers (PEO-PPO; Symperonic® F38, MW 4.8 k; Synperonic® F68, MW 8.35 k; Synperonic F88, MW 11.8 k; ICI Iberica, Spain), bovine serum albumin (BSA, Sigma), tripolyphosphate (TPP, Sigma) and tetanus toxoid antigen (TT, 8,500 Lf/mL, Massachusetts Public Health Biologic Laboratories, MPHBL) and diphtheria toxoid (DT, 540 Lf/mL, MPHBL).

2.2
Equipment and Procedures

2.2.1
Stoichiometry Evaluation in a Batch System

This system consisted of a needle (gauge #26) connected to a 5-mL syringe inserted into an ultrasonic hollow titanium probe with a 1.85 mm ID conical tip. The probe was connected to a converter (transducer) and a power generator (Ultrasonic Nebulizer XL 6040, Misonix Inc., Farmingdale, NY) with an output power and frequency of 30 watts and 40 kHz, respectively. Anionic polymer solution was introduced into the syringe and slowly extruded (by means of controlled air pressure) through the needle and the probe tip (in the form of fine mist) into the air above a cationic solution placed in a beaker (Fig. 1). A 50-mL beaker (receiving bath) was swirled vigorously by hand during the reaction. Usually, 1–2 mL of anionic solution was captured (from a 3-inch distance) in 20 mL of cationic solution (anion/cation ratio 1/20 to 1/10) in 1–2 minutes. The extrusion pressure was typically 3 psig and the flow rate 1 mL/min. In a standard case, the anionic solution (see below) was used as a droplet-forming internal phase mixture (core polymers, C), and the cationic solution (below) as a receiving bath mixture (corona/shell polymers, S). The reverse system included cationic core polymers and anionic corona polymers. The core polymer was introduced and captured into the corona solution in the form of a mist. Nanoparticles were produced immediately upon contact of the extruded liquid with the receiving bath.

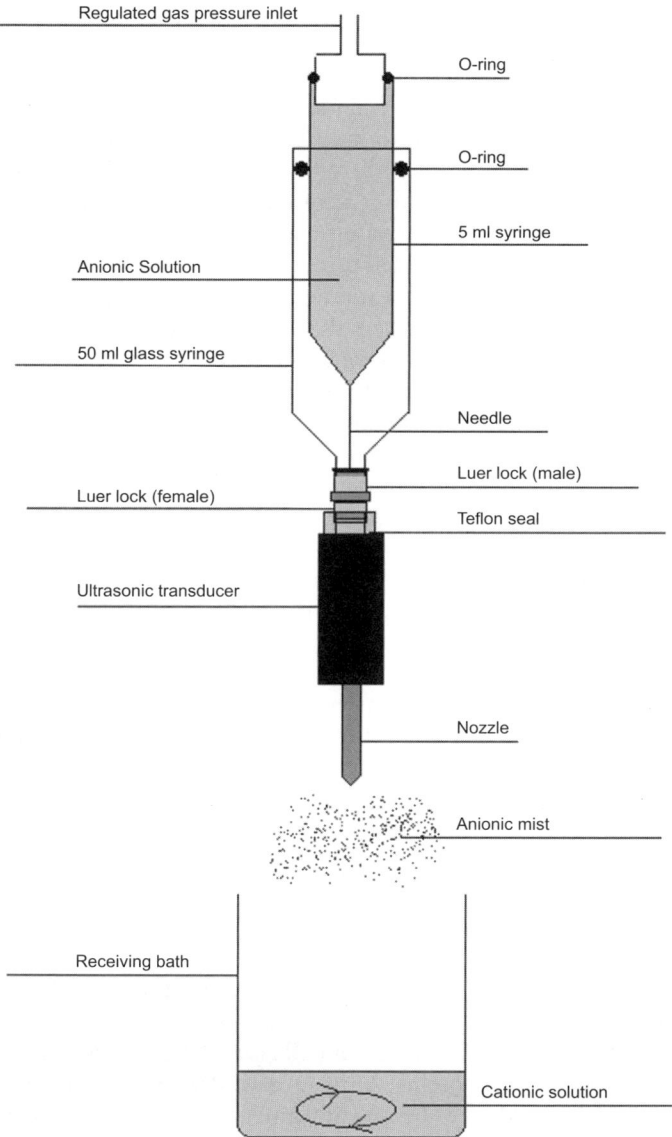

Fig. 1. Diagram of a batch set-up. The cationic bath solution is initially purely cationic, however, when the production begins by introducing an atomized anionic polymer solution it will in part be converted into the final product, a suspension containing nanoparticles. The mist was captured from a 3-inch distance

2.2.2
Stoichiometry Evaluation in a Continuous System

This system was a modification of the batch system made by introducing two inflow lines (anionic and cationic polymer solutions) and one overflow line (not shown). The latter line was added to keep the receiving bath volume constant (50 mL). The reactor contents were stirred with a magnetic bar. This system is effectively a continuous stirred tank reactor (CSTR), a standard piece of equipment in the chemical industry. The anionic solution was again dispensed in the form of a spray with the flow regulated through the syringe/ultrasonic transducer system. The anion/cation flow rate was adjusted to be in the range of 0.5/20 to 1.5/20. The data were collected when the liquid volume changeover was 3–4 volumes, based on the calculated average residence time.

Once the particles were produced by either method, they were analyzed for size and charge in the reaction mixture directly or after separation according to the procedures described in the next section.

For stoichiometry evaluation, nanoparticles were produced via a batch process, product was isolated, freeze-dried, and sent for elemental analysis. Several batches of particles were made, pooled and processed as described below. The pooled volume was centrifuged in 30-mL plastic centrifuge tubes (Beckman Model L5–50 Ultracentrifuge, Rotor Type 60 Ti) for twenty minutes at 15,000 rpm. The supernatant fluid was discarded, the sediment resuspended in 1 mL of sterile water, vortexed for 10 seconds (to remove the nanoparticles from the wall of the tube) and, finally, 25 mL of water was added and the suspension centrifuged a second time. The second centrifugation step was repeated. Following the third centrifugation the product was re-suspended in a minimum volume of water (0.8 mL per centrifuge tube) and vortexed again. All portions were then pooled together.

2.2.3
Preparation of Nanoparticles Loaded with Model Proteins for Release Experiments

Three proteins were used for initial studies: OVA, CC and bFGF. The core solution contained 0.05 % HV and 0.05 % CS whereas the shell solution consisted of 0.05 % SP, 0.15 % PMCG, 0.05 % calcium chloride and 1 % F-68. The core solution also contained model proteins, 0.5 % OVA or 0.02 % CC when loaded. The C/S ratio was 2/20 (mL/mL). For bFGF entrapment, the same chemistry as above was used. This is only possible as long as the amount of bFGF was small, a requirement easily satisfied for growth factors (therapeutic doses are usually several orders in magnitude lower compared with standard drugs). Otherwise, a protein such as bFGF, a basic molecule under physiologic conditions, would precipitate out while introducing it into the anionic solution.

For cross-linking, the core solution also contained PDA. The mass ratio PDA/OVA was 0.1 and 0.2 for 50-CL, 100-CL products, respectively, and PDA/CC was 0.025 and 0.05 for 50-CL and 100-CL products, respectively. The term CL denotes cross-linked product. For PDA-loaded particles, following their centrifugation, they were resuspended in a bicarbonate buffer (pH 8.3), incubated for 15 min at 37 °C to form Schiff-base complex and subsequently washed again.

2.2.4
Preparation of Nanoparticulate Vaccine System Loaded with Model Antigens

Particle chemistries were generated using several core/shell systems (Table 4), listing corona (and antigen) and shell solution composition, their ratio at production and dose used in mice. All particles were positively charged. The preparation of antigen (TT, DT)-loaded CT/TPP nanoparticles was described by Calvo et al. [14].

2.2.5
Preparation of Nanoparticulate System Loaded with GMS-Polymer Conjugate

Gentamycin sulfate, GMS (Sigma), is a small molecular weight drug compound (MW of 710 Da), representing a class of cationically charged drugs. To prepare the conjugate [16], 50 mg of gentamycin sulfate were dissolved in 2.5 mL water and then 0.53 M $NaHCO_3$ was added to make the solution alkaline. The final volume was 3.5 mL and the bicarbonate concentration was 0.1 M. 100 mg AKM-1510 maleic anhydride-PEG polymer (Shearwater Polymers, Huntsville, AL; average MW 14 kDa) were then added to the GMS solution and this solution was kept at 4 °C under stirring. The product was dialyzed against water (800 mL, twice) at 4 °C. The antimicrobial activity of this product was estimated by *in vitro* disc diffusion assay using a test organism on agar plates (*Bacillus subtilis*) [17] and compared with free GMS.

The nanoparticles were generated using a droplet-forming polyanionic solution composed of 0.025 % HV, 0.025 % CS in water (and GMS) and corona-forming polycationic solution composed of 0.05 % SP, 0.065 % PMCG, 0.05 % calcium chloride and 1 % F-68 Pluronic in water. The working anionic solution consisted of 5.4 mL of the above anionic solution and 0.6 mL of dialyzed conjugate mentioned above. The C/S ratio was 2/20 (mL/mL). The particles, which instantaneously formed, were allowed to react for 1 hour while stirring and then collected by centrifugation at 15,000 g for 20 min. A 1-mL aliquot of the resulting suspension contained about 5.7 mg GMS incorporated into nanoparticles. The GMS release was assessed via an efflux method in vitro using 5 mL test buffer (PBS) on a shaker. Samples were withdrawn at different time intervals and the amount withdrawn was replaced with a fresh buffer. GMS was measured by disc diffusion assay.

Table 4. Experimental vaccine systems

System #	Core (C) components (wt. %) (± antigen)	Shell (S) components (wt. %)	C/S ratio	Dose mg/mice	Results (Ab titre)	LE, % (2 batches)
1-OVA	0.1 % TPP, 0.1 % X52 (0.3 % OVA)	0.05% CT, 0.1% $CaCl_2$, 1 % F-68	1.8/20	0.5	Negative (SQ)	15.9; 14.9
2-OVA	0.025 % HV, 0.025 % GL, 0.0125 % HMP (0.4 % OVA)	0.075 % PMCG, 0.05 % $CaCl_2$, 1 % F-68	2.0/20	2	Fig. 22 (SQ)	45.0; 42.8
3-TT	0.05 % HV, 0.025 % GL, 0.0125 % HMP (0.012 % TT)	0.075 % PMCG, 0.05 % $CaCl_2$, 1% F-68	3.5/20	0.025	Negative (OR)	1.8; 1.6
4-TT-CL	0.025 % HV, 0.025 % CS, 0.0014 % PDA (0.012 % TT)	0.05 % SP, 0.065 % PMCG, 0.05 % $CaCl_2$, 1 % F-68	2/20	0.025	Fig. 24 (SQ)	2.0; 1.8
5-OVA-CL	0.025 % HV, 0.025 % CS, 0.014 % PDA (0.67 % OVA)	0.065 % PMCG, 0.05 % $CaCl_2$, 1 % F-68	2/20	0.5	Fig. 23 (OR); SQ also positive, similar to Fig. 24	20.3; 25
6-OVA	0.025 % HV, 0.025 % CS (0.5 % OVA)	0.05 % SP, 0.065 % PMCG, 0.05 % $CaCl_2$, 1 % F-68	2.2/20	0.5	Negative (OR)	10.3; 14

2.2.6
Aseptic Operation

For aseptic preparation, the entire equipment assembly [18] was exposed to ethylene oxide sterilization, followed by an overnight degassing in a standard hospital gas sterilization autoclave. The gas input for the extrusion of the anionic liquid was passed through a 0.2 µm Acrodisc hydrophobic filter (Fisher, Pittsburgh, PA). The actual processing was carried out in a sterile laminar flow hood, as well as the product pellet handling and washing steps. Sterile preparations were produced for use in *in vivo* and long-term *in vitro* release experiments.

2.2.7
Systems Used for Comparison

Because of extensive references to other papers [14, 15] we report on the procedure for preparation of CT/TPP nanoparticles loaded with BSA. Only one optimized CT/TPP system is mentioned, with variable amounts of entrapped surfactants. The reactant solutions contained 1.25 mg/mL CT (containing varying amount of PEO/PPO, 0–200 mg/ml) and 0.28 mg/mL TPP. Nanoparticles were formed spontaneously upon incorporation of 2 mL of the TPP aqueous solution to 5 mL of the CT acidic solution (pH 5.0) under magnetic stirring. The BSA was incorporated by an equilibrium partitioning method or by incubating the nanoparticles with an acidic solution containing various concentrations of BSA or by including BSA in either CT or TPP solution before preparing nanoparticles. Nanoparticles were isolated by centrifugation (40,000 g, 10 °C, 30 min in the presence of 5 % trehalose) and resuspended in water by manual shaking. It appeared that the process of formation of primary droplets, as employed in our method, might not be necessary. A simple mixing of two "reacting" solutions, according to Calvo et al. [14, 15], resulted in nanoparticles, provided certain concentration range was applied. According to our experiments (not presented here), some chemistries may not result in satisfactory nanoparticle production by a simple mixing process. Many chemistries generated a mixed product composed of aggregates and nanoparticles, a non-satisfactory situation.

2.3
Encapsulation Efficiency and Drug Loading

The entrapment efficiency (EE) was defined as a percentage of protein mass recovered in the final product (after washing) to the originally provided protein mass (in the anionic solution). The concentrations of protein in both streams were evaluated using the BCA method [19], following product dissolution after heating of nanoparticles at 60 °C in an alkali solution for 2 minutes. Drug loading or loading efficiency (LE) was defined as the percentage of protein mass to the total mass of the final product. Again, the BCA method was employed. The drug loading capacity then becomes the maximal amount trapped in the product (in %) for a particular nanoparticulate chemistry and drug.

2.4
Product Analysis (for Stoichiometry)

The final product was freeze-dried (Labconco Freeze Dryer, Fisher, Pittsburgh, PA), and then both unloaded and loaded samples were shipped for elemental analysis (Quantitative Technologies, Inc., QTI, Whitehouse, NJ). The C, H, N, P and S analysis is reported with a 0.3 % precision (absolute). The calcium analysis was performed via atomic absorption spectroscopy using inductively coupled plasma and a graphite furnace. All analyses were carried out only after moisture was removed via a brief heating at the QTI site, even though the samples were shipped in a dry state. The product and component moisture analysis was assessed by a standard method (heating a sample for 2–3 hours at 105 °C in a oven). The product content of oxygen in cases when S was present was estimated by difference.

The principle of mass conservation, on a macroscopic basis, using equations representing the conservation of total mass, the conservation of each chemical element and the mass balance for each of the molecular species was applied. Chemical reactions were taken into account in these equations. The stoichiometric coefficients of individual reactants were evaluated using Microsoft Excel software by expressing the equations in matrix notation.

2.5
Measurement of the Primary Droplets

Attempts were made to photograph the anionic droplets and measure their size using a digital camera with flash exposure (DALSA CAD1 with LabVIEW software).

2.6
Intermediate and Product Particle Characterization

The size and charge analysis was done using a Coulter DELSA 440SX (Coulter Beckman Corp., Miami, FL). This particular instrument measured the size distribution on the basis of photon correlation spectrometry (PCS) and was limited to particle diameters between 0.02 µm and 3 µm. Measurements were taken at four different angles simultaneously with 256-channel resolution each. Comparison of the spectra allowed for the detection of very small particles. The zeta potential was assessed on the basis of electrophoretic mobility (laser Doppler anemometry, LDA). This was defined as the particle velocity per unit of applied electrical field, with units usually given as $\mu m\ s^{-1}/V\ cm^{-1}$, while zeta potential is defined as the electrical potential between the bulk solution and the shear plane around the particle, with units usually given as millivolts (mV). Calvo et al. [14, 15] employed similar methods for size and charge measurements, except that the instrument used was Zetasizer® III (Malvern Instruments, UK).

To monitor nanoparticle swelling in salt environment, we employed laser diffraction Mastersizer Micro Particle Analyzer MAF5000 (Malvern) with a dynamic range of 0.3 to 300 µm. This instrument utilizes Mie scattering algorithm with the Fraunhofer approximation.

2.7
Product Stability

Once the particles are formulated they must pass a rigorous test of stability. Colloidal stability (non-aggregation) was performed at a moderate salt concentration (0.9 % w/w, to simulate the physiologic environment) at pH 2–8.5 using centrifugation (for their recovery) to define their practical usefulness. As a control, a suspension of nanoparticles in water (or neutral buffer) was used. The pH range selected above allows for NP applications for some extreme tissue environments. The stability tests were performed at 4 °C, 20 °C and 37 °C over 1–4 week periods. The centrifugation step was also important. As the particles are compacted during the centrifugation step, the inter-particle distances and shear-induced zeta potential distance become shorter and the tendency to aggregate becomes more pronounced. A stable turbidity reading as compared to the initial condition prior to the exposure (as measured in a spectrophotometer supplemented by visual observation) signified a stable suspension, which is the end-point criterion. The particle size was also used as another end-point criterion. Since it is not clear which polymeric vehicle chemistry would pass all the tests, we normally screened several different chemistries to enhance our chance of success. Typically, the entrapped substance exhibited some effect on particle stability, and this effect was difficult to predict. A special stability test of a single-pair polymer complex vs a multiple-component complex will be explained in the Results Section.

2.8
Product Freeze-Drying

Aliquots of nanoparticulate suspension were frozen at –20 °C in the presence of 0, 5, 10, 15, 20, 25 and 30 % trehalose. Samples were freeze-dried in a Virtis Co. (model BT 6.6 × L, Gardiner, NJ) system under the following conditions: a primary drying step for 24 hours at –30 °C and secondary drying step until the temperature gradually rose to 20 °C. The particles were then resuspended in distilled water (to return to starting trehalose concentration), and their size and charge were measured.

2.9
Release Measurement and Permeability Control

We employed two different methods to adjust the release kinetics. The polymeric matrix itself served as a slowly sequestering chemical container. In addition, we employed a reactive step using relatively non-immunogenic PDA [16,20], applied into the core polymeric formulation. A 30-min incubation of recovered nanoparticles at pH 8.3 (Hepes buffer) provided cross-linking conditions. This step was then followed by another wash by water. The final product was assumed to feature Schiff base linkage between the carboxyl group of polysaccharides of the polymeric mixture and the amino groups of proteins. No reduction step was applied to generate a covalent bond. The *in vitro* release rates were evaluated in different buffers (details are in respective figures), at two different temperatures (20 °C and 37 °C). The release tests were completed using three proteins: OVA, CC and bFGF. OVA was assayed via the BCA method [19] and CC spectrophotometrically at 408 nm. At different time intervals, the solution (buffer) containing the released protein was removed for quantitation and replaced with new buffer. Typically, daily replacement was used. bFGF release was measured via an ELISA kit (Quantikine; R & D Systems, Minneapolis, MN), following the manufacturer's protocol. The biological activity of released bFGF was assayed via a proliferation assay (CellTiter 96 AQ_{ueous} One Solution Cell Proliferation Assay), based on formazan formation using MTS tetrazolium compound (TB 245, Promega, Madison, WI). NIH 3T3 fibroblasts, plated at a density of 2000 cells per well (96-well plate) were serum-starved for 16 hours in Optimem medium prior to the 3–5 day assay.

For detailed bFGF release studies, the protein was labeled (^{125}I) with the help of IODO-BEADS iodination reagent (Pierce, Rockford, IL) [21]. An aqueous solution of bFGF at 6.6 mg/mL (100 µL) was added into 200 µL of reaction buffer (20 mM sodium phosphate pH 7.5, 150 mM NaCl) containing one polystyrene bead with immobilized iodination reagent and 2.5 µ of $Na^{125}I$ solution (13674 MBq/mL in 0.1 N NaOH, NEN, Perkin Elmer, Shelton, CT). After 5 minutes at room temperature the reaction was stopped by removing the solution from the reaction vessel and the labeled protein was passed trough a PD-10 column (Amersham, Arlington Heights, IL) in order to remove uncoupled, free ^{125}I molecules. The labeled protein was eluted by adding the reaction buffer containing 0.1 % BSA to the column. Aliquots (5 µL) of all fractions (500 µL) were counted for a gamma radiation with a gamma counter to identify the labeled protein.

2.10
Compatibility and Animal Data

The tissue reactivity of nanoparticles was tested following intravenous injections in C57Bl6/DBA hybrid mice aged between 10–12 weeks. Different quantities of unloaded nanoparticles ranging from 3 to 12 mg dry weight per mice were used. In addition, mice were injected in the tail vein with 200 µL of three different nanoparticle suspensions containing respectively 3 mg, 6 mg and 12 mg of material using a 0.3 mL syringe and a 28 G needle. The study was based on the use of 5 mice per group per dose. Previously [22], we have established that it is meaningless to test individual polymeric components for their cytotoxicity *in vitro*. Polycations in particular exhibited substantial cytotoxicity over the practical range of concentrations. However, when combined in a complex (as is the case of nanoparticulate formulation), their cytotoxic effects were substantially reduced if not absent. Following the injections, 24 hours later the animals were sacrificed and their organs (lung, spleen, heart, kidney, liver) were excised, rinsed quickly with cold water to remove surface blood, fixed in 10 % neutral buffered formalin for 48 hours, dehydrated, and embedded in paraffin in preparation for sectioning.

Representative sections (6 µm) of each organ were stained with Masson's trichrome and hematoxylin/eosin.

2.11
Immunizations and Immunoassays

Mucosal and systemic antibody responses were measured following oral (OR) and SQ immunization by a nanoparticulate-formulated protein. For oral immunizations, C57BL/6 female mice, 6–8 weeks old, were used in groups of 10 animals. Animals were immunized by gavage using a dose volume of 500 µL. Experimental formulations were prepared so each dose contained a certain load of protein (Table 4). Formulations tested included soluble protein, protein in the nanoparticulate form, protein mixed with empty nanoparticles and protein formulated with CRL-1005 non-ionic block copolymer. The immunization protocol included two immunizations one month apart. Blood, feces and saliva were collected from immunized mice on study days 21, 28, 49 and 56 and stored at –70 °C until tested.

The total antibody titers and antibody isotypes were determined using sera, saliva or PBS solubilized fecal material from immunized mice in an ELISA assay in a 96-well microplate format. Protein in the nanoparticulate form was used as the ELISA test antigen and test wells without protein were used to determine background or non-specific levels of antibody binding. Mouse sera were tested in duplicate using a four-step $\log_3$ titration starting at 1/56 while saliva and the solubilized feces were tested at dilutions of 1/5 and 1/20. An anti-globulin reagent specific for both the light and heavy chains of murine immunoglobulins (conjugated directly to horseradish peroxidase, Sigma) was used for measurement of all antibody isotypes. The colorimetric substrate used was 3,3',5,5'-tetramethylbenzidine (Kirkegaard and Perry Labs, Gaithersburg, MD) and the absorbance of individual reactions was read at 450 nm using a 96 well microplate reader. Endpoint titers were determined using a reference antiserum and a standard curve.

3
Results

3.1
Rationale

The research discussed in this review involves polymeric nanoparticles that are constructed of a mixture of natural, as opposed to synthetic, polymers. Water is used as a solvent in this technology as contrasted to organic-based technologies. Therefore, there is no chance of release of toxic solvents during biodegradation. Trace amounts of hydrophobic toxic organic substances could accumulate over time in adipose tissue, causing potential health risks. The nanoparticles described in this review are composed of natural polymers making them excellent candidates for drug carriers, as they do not introduce health risks from toxic materials.

3.2
Nanoparticle Assembly and Structure

The polymeric nanoparticles described in this paper are produced by electrostatic interaction between anionic and cationic solutes (polymeric complexing). The anionic droplet contains approximately a 1/1000 mass ratio of polyanion to water. When the anionic and cationic solutions are brought together, the water is partially expelled, and the nanoparticle is formed as a result of hydrophobic forces. Such nanoparticles are typically insoluble in water. By measuring the size of the primary anionic droplet and calculating the mass ratio of anion to water, it is possible to simply calculate an approximate size of the final product. This option was explored briefly as a method of controlling size. The primary droplet size was measured close to the detection range of the equipment used (the smallest size detectable was 10 µm). It is noted that under the assumption that all anionic components are integrated into the final product and the primary droplet size does not aggregate as a result of processing (centrifugation and washing steps) the expected size would be 10 nm. We typically observed, however, a minimal size ten times larger. That implies that as a result of inotropic complex formation from primary solutions a large amount of water is rejected (but not all) due to the assembly process (charge neutralization and hydrophobicity raise).

It is assumed that the nanoparticles consist of two main parts: the inner core and the semi-permeable membrane shell (Fig. 2). The inner core usually holds the desired drug. The semi-permeable membrane, in part, controls drug leakage into the body. The particle is porous, and the administered drug continually diffuses out through these pores. Most often, nanoparticles have an anionic core with a cationic shell, which is the desired formulation for many drug delivery applications, which have anionically charged (or uncharged) drugs. For cationically charged drugs, however, a reverse system is also possible with a cationic core and an anionic shell. A few examples of some successful chemistries are listed in Table 2.

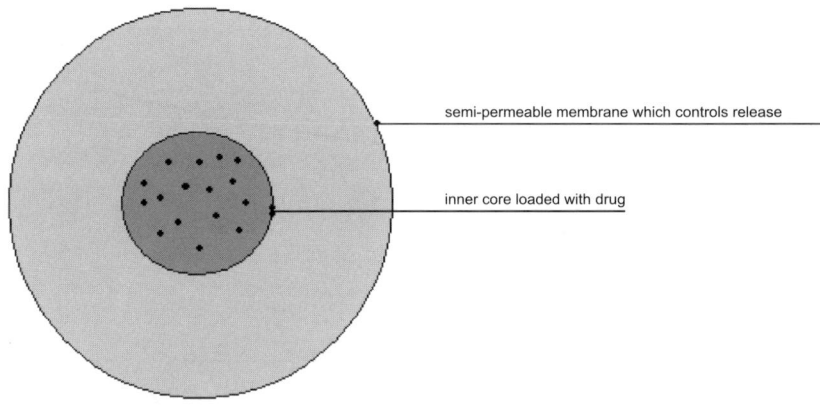

Fig. 2. Diagram of nanoparticle structure (core-shell morphology)

Many products with different chemistries [a combination of at least two anionic polymers in the droplet-forming phase and one (or two) cationic polymer(s) plus a small inorganic ion as a core-forming phase] have been produced. However, not every experiment yields a usable product. Many combinations lead to an undesirable aggregated product. In all cases, the suitable concentration of steric stabilizer (F-68) must be found empirically. The reverse chemistry is more difficult to formulate.

For both batch and continuous production, the lower limits of added anion volume were chosen to provide sufficient mass of product for good measurements to be taken. The upper limit was selected to avoid charge neutralization [23], since the nanoparticles tend to aggregate under these conditions, making size and charge measurement impossible. Aggregates are useless as practical nanoparticles.

The model protein chosen for the loaded nanoparticles was ovalbumin (OVA), a weak antigen.

3.3
Entrapment and Loading Efficiencies

Entrapment efficiency depended on the concentration of the core polymeric solution. No attempts have been made to optimize this parameter. Currently, the entrapment efficiencies for proteins are in the 5–20 % range, and the loading efficiencies are between 10–50 %. Figs. 3 and 4 present batch data for variable amounts of OVA in the reaction mixture (System #2). In either case no saturation was reached.

For a binary system (one polycation and one polyanion on each interaction side), CT/TPP, Calvo et al. [15] described very high entrapment efficiencies from 20 to 92 %. CT/TPP PEO or PEO/PPO polymers were co-entrapped within the final product. Nanoparticles were formed spontaneously by simple mixture of

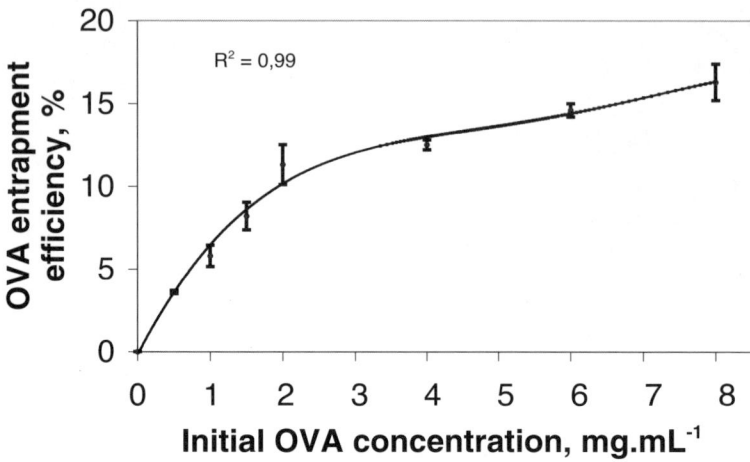

Fig. 3. Effect of initial ovalbumin concentration on entrapment efficiency in a batch system. Nanoparticulate chemistry (System #2) was as follows: Core: 0.05% HV, 0.05% CS, variable amount of OVA, 1–8%; Shell: 0.05% SP, 0.075% PMCG, 0.05% $CaCl_2$, 1% F-68. C/S ratio = 2/20 (mL/mL). Data are presented as the mean ± SD ($n = 4$)

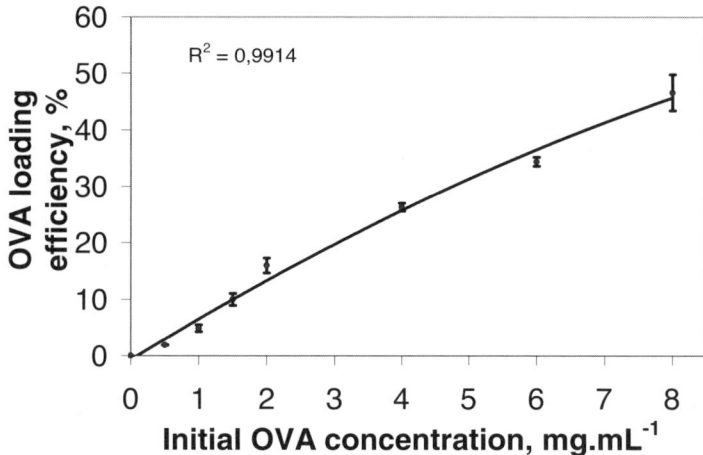

Fig. 4. Effect of initial ovalbumin concentration on ovalbumin loading efficiency in a batch system. Nanoparticulate chemistry (System #2) was as in Fig. 3. Data are presented as the mean ± SD ($n = 4$)

two solutions with stirring. These authors systematically studied the effects of pH and BSA concentration on the entrapment and loading efficiency. The total amount of protein entrapped per weight of nanoparticles (loading efficiency) was related to the BSA concentration (Fig. 5). The maximum entrapment efficiency was obtained when nanoparticles were prepared at pH 5. At this pH BSA is negatively charged (pI is 4.8). The EE also depends (in an inverse fashion) on

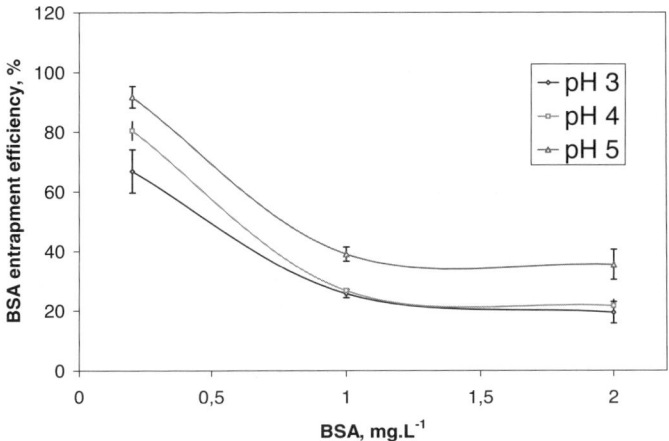

Fig. 5. Effect of initial BSA concentration and pH of CT solution on entrapment efficiency (mean ± SD, $n = 3$) for CT/TPP nanoparticles (reproduced from Calvo et al. 1997, with permission of Plenum Press; [15])

the concentration of BSA. The highest EE were achieved with the lowest BSA concentration. According to the authors, the main mechanism of the particle loading/release was protein-polysaccharide (chitosan) electrostatic interaction. The results of Calvo on the EE (Fig. 5) are in sharp contrast to the behavior of OVA in System #2 (Fig. 3). At this moment, we do not have any explanation for the nature of this discrepancy.

3.4
Size and Charge Data

The data from the size and charge measurements of nanoparticles produced in both batch and continuous systems are presented in Figs. 6–9, respectively. Each data point represents the average of 3 to 5 measurements along with the standard deviation.

Separate experiments revealed that the size/charge values were similar when done in the reaction system or after the particles were isolated and washed. Indeed, the presence of the reaction mixture provided a high degree of stability to the product. The produced nanoparticles were extremely stable (for many months) in the presence of excess cationic solution. As the ratio of anion to cation increased, the charge on the particle periphery decreased. This confirmed our hypothesis that the charge on the particle periphery would be greatly influenced by the stoichiometry between the reaction components from which the particle was derived. As the amount of anion added to the solution increased, the charge (positive) on the periphery of the resulting nanoparticles decreased. Continuing in this fashion would lead to neutrally charged particle chemistry and to product aggregation.

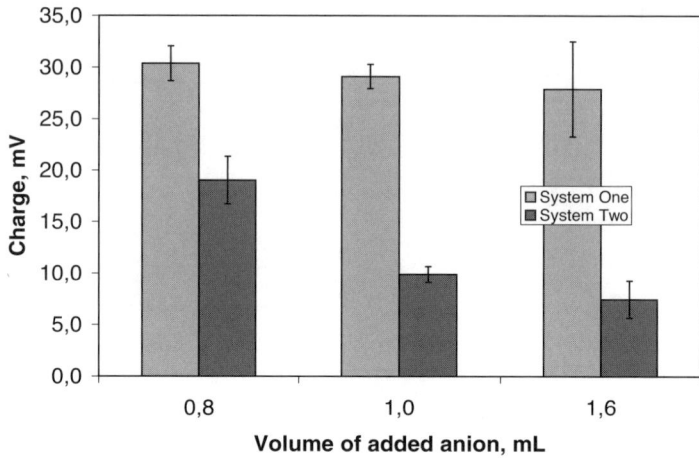

Fig. 6. Dependence of nanoparticle charge on volume of added anion in a batch system. Nanoparticulate chemistries were as follows – System #1: Core: 0.1 % TPP, 0.1 % X-52; Shell: 0.05 % CT, 0.1 % $CaCl_2$, 1 % F-68. System #2: Core: 0.05 % HV, 0.05 % CS; Shell: 0.05 % SP, 0.075 % PMCG, 0.05 % $CaCl_2$, 1 % F-68. Data are presented as the mean ± SD ($n = 3$)

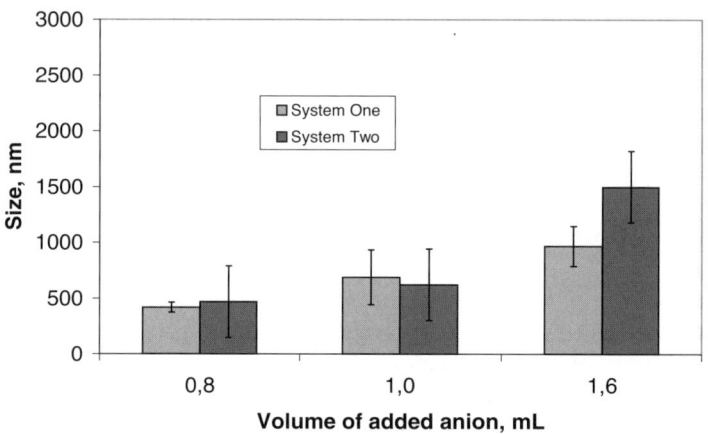

Fig. 7. Dependence of nanoparticle size on volume of added anion in a batch system. Nanoparticulate chemistry was as follows: System #1 as in Fig. 5, System #2 as in Fig. 5. Data are presented as the mean ± SD ($n = 3$)

Regarding size, it was apparent that particle size increased with the anion to cation ratio. The scatter of data, however, indicated that batch processing was generally quite unreliable.

Data on charge and size measurements supported our visual observations on particle stability, particularly for larger volumes. A sudden (and even precipitous) aggregation is observed when the anion volume added to the system surpasses 1.6 mL for System #1 and 1.4 mL for System #2.

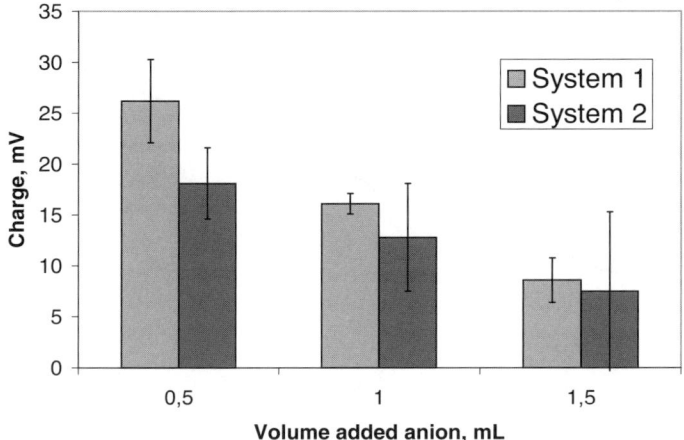

Fig. 8. Dependence of nanoparticle charge on volume of added anion in a continuous system. Nanoparticulate chemistry was as follows: System #1 as in Fig. 5, System #2 as in Fig. 5. Data are presented as the mean ± SD ($n = 4$)

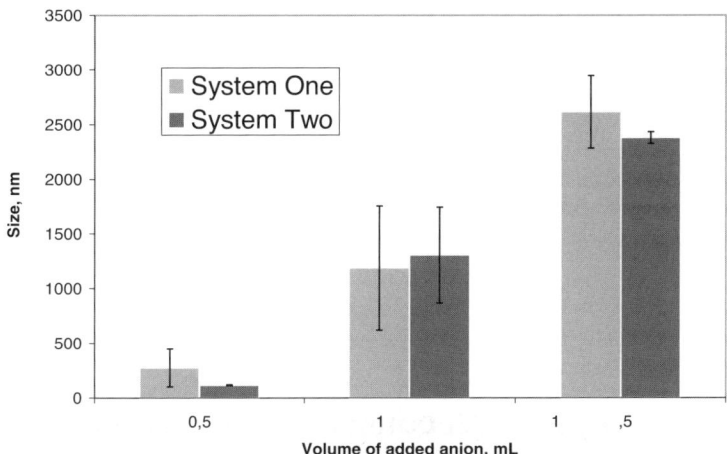

Fig. 9. Dependence of size on volume of added anion in a continuous system. Nanoparticulate chemistry was as follows: System #1 as in Fig. 5, System #2 as in Fig. 5. Data are presented as the mean ± SD ($n = 4$)

In a continuous system, it was seen that the same trend of size and charge data occurred. The continuous data displayed much clearer trends, however. Continuous processing is thus the preferred method of nanoparticle production, enabling a better quality control.

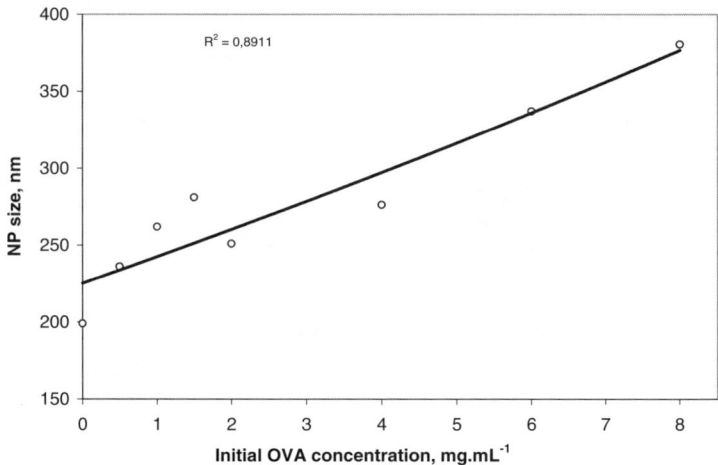

Fig. 10. Effect of initial ovalbumin concentration on product size in a batch system. Nanoparticulate chemistry (System #2) was as in Fig. 3. Data are presented as the mean ± SD ($n = 4$)

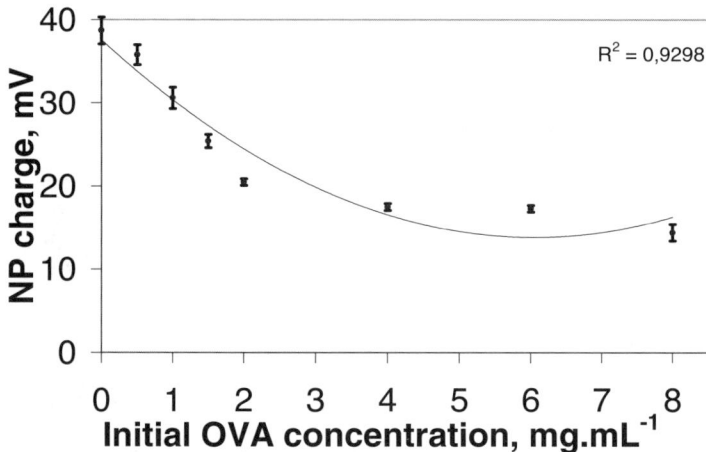

Fig. 11. Effect of initial ovalbumin concentration on product charge density in a batch system. Nanoparticulate chemistry (System #2) was as in Fig. 3. Data are presented as the mean ± SD ($n = 4$)

Separate batch data on size and charge (Figs. 10 and 11) were collected to complement the results presented in Figs. 3 and 4. They confirm the findings obtained from batch and continuous systems (Figs. 6–9).

It was interesting to compare the above size and charge data with those obtained by other authors. Calvo et al. [14] collected such data for their single component CT/TPP system. Fig. 12 shows the effect of CT and TPP concentrations on product size (in the absence of PEG or PEO/PPO surfactants). Figs. 13 and 14

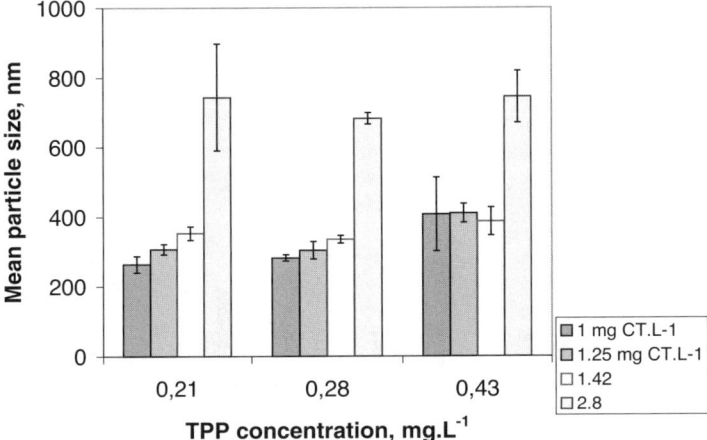

Fig. 12. Mean particle size (in nanometers ± SD, $n = 4$) of chitosan/TPP nanoparticles as affected by reagent concentrations (reproduced from Calvo et al. 1977, with permission from Wiley) [14]

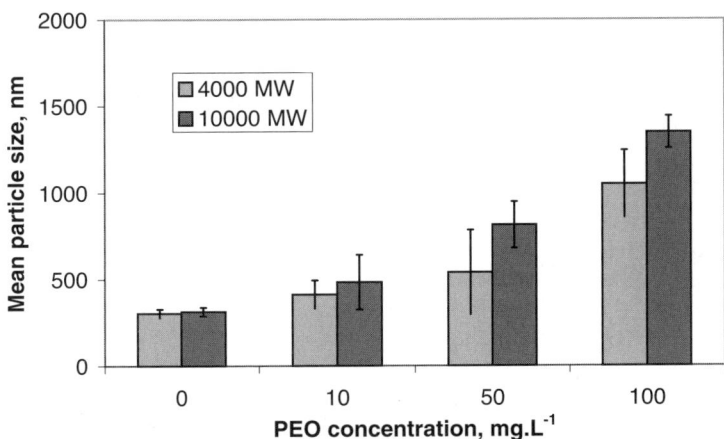

Fig. 13. Mean particle size (in nanometers ± SD, $n = 4$) of chitosan/TPP nanoparticles as affected by PEO initial concentration in the chitosan solution and PEO molecular weight (reproduced from Calvo et al. 1997, with permission of Wiley) [14]

show the effect of surfactants on size and charge, respectively. It appears that size increased with the amount of surfactants (and their MW) added into the reaction mixture. The zeta potential was lower with the increased amount of surfactants added, indicating their increased incorporation into the product (Fig. 15). The correlations observed for the CT/TPP system may have a relevance to our formulation.

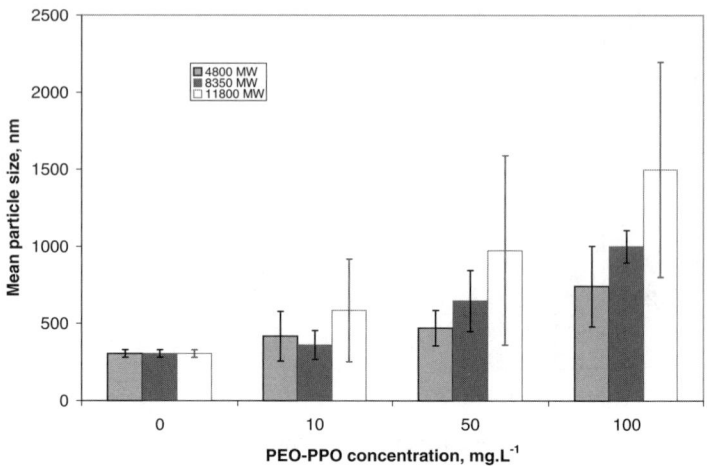

Fig. 14. Mean particle size (in nanometers ± SD, $n = 4$) of chitosan/TPP nanoparticles as affected by PEO-PPO initial concentration in the chisosan solution and PEO-PPO molecular weight (reproduced from Calvo et al. 1997, with permission of Wiley) [14]

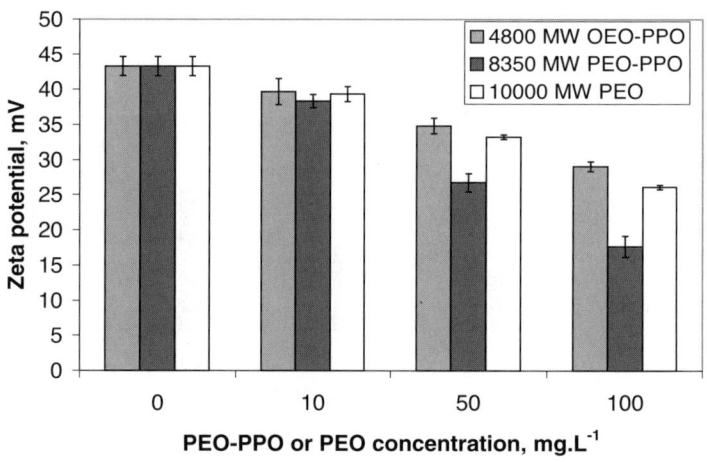

Fig. 15. Zeta potential (in mV ± SD, $n = 4$) of CT/PEO-PPO and CT/PEO nanoparticles as affected by concentration in the initial chitosan solution and molecular weight of surfactants (reproduced from Calvo et al. 1997, with permission of Wiley) [14]

3.5
System Stoichiometry

The polymer chemistry (the choice of polymers in two polymeric mixtures) was selected to allow identification of individual contributions to the product composition (System #1). Generic equations can be proposed to provide a framework for experimentation and calculations, for reactions involving both unloaded and loaded nanoparticles:

a TPP + b X-52 + c Chitosan + d $CaCl_2$ + e F-68 → Product   (a)

a TPP + b X-52 + c Chitosan + d $CaCl_2$ + e F-68 + f OVA → Product   (b)

In order to determine the amount of each compound that was incorporated into the final nanoparticle product, calculations were performed around a signature element that was present in each particular compound. P was used as a signature element for TPP, S for X-52, Ca for calcium chloride, and N for chitosan. By assaying the content of those distinct elements in the individual input "reactants" and in the final product, we assigned stoichiometric coefficients to individual reactants. The data were first corrected for moisture (water loss). However, no corrections for the moisture content were used for TPP and $CaCl_2$ since they were available in dry form. Pluronic F-68 melted and decomposed upon heating (therefore, it was assumed to contain no moisture). The weight percent results of the elemental analysis in terms of carbon, hydrogen, nitrogen, sulfur, phosphorous and calcium content were then converted into moles. The oxygen content was then calculated by difference. A degree-of-freedom analysis showed that for unloaded nanoparticles there were four input variables (four distinct elements) and five reactants in the overall reaction scheme. The fifth reactant with no distinct elements was Pluronic F-68. For that reactant a mole/mass balance around the hydrogen atom was used.

In order to assign stoichiometric contributions of both chitosan and ovalbumin (both contain nitrogen) to the final product for loaded nanoparticles, the amount of protein nitrogen contributed by ovalbumin was assessed by a spec-

Table 5. Calculated values of stoichiometric coefficients as dependent on the anion/cation ratio (C/S). Data were obtained from 15 pooled samples; the last two lines show variability between two separate runs

Ratio	TPP	X-52	Chitosan	$CaCl_2$	F-68 (H)	OVA
1.4/20	0.031	0.124	0.29	0.002	0.02	–
1.6/20	0.012	0.090	0.29	0.002	0.03	–
1.8/20	0.008	0.127	0.28	0.005	0.01	–
2.0/20	0.112	0.093	0.27	0.003	0.03	–
2.0/20 #1	0.035	0.068	0.22	0.001	0.01	0.34
2.0/20 #2	0.029	0.073	0.28	0.006	–0.003	0.32

trophotometric method (BCA; [19]). The contribution of F-68 was again analyzed by means of a hydrogen balance. Table 5 presents the calculated values of moles of each reactant incorporated into the final product. In the case of the last entry in Table 5, a negative value of the coefficient was noted for F-68. The reasons for this discrepancy could be several:
1) overestimates of the other compounds;
2) a low content of F-68 in the final product; or
3) experimental errors on the part of all analyses involved.

A low content of F-68 in the final product was not surprising. Uncharged chains of F-68 are merely mechanically entrapped during the nanoparticle assembly process, which involves charged molecules of the reactants. The inclusion of F-68, however, is crucial for particle steric stabilization.

3.6
Product Stability

Stability of nanoparticles in the reaction mixture in these systems was unprecedented. The unreacted cations appear to stabilize the suspension. Stability of the isolated nanoparticles in water (non-loaded) was also very good (no change in particle size up to three weeks at 4 °C; size 226 nm, SD 28.2, $n = 7$; chemistry used was System #2, Table 1). Substantial swelling then occurs. In saline (0.9 % w/w) the same particles exhibit a rapid increase in size over six hours, from 0.2 nm to 2.6 µm. We have not systematically tested swelling of other nanoparticulate chemistries.

Simple binary interactions were not enough to achieve a long-term stability [24, 25]. Table 6 presents relative stabilities of binary and quaternary complex structures. This table qualitatively illustrates the concept that multiple interactions are necessary in order to stabilize the final product. Both binary and some ternary interactions in Table 7 showed instability throughout the matrix. This behavior was very common. Matrix stability was normally evaluated in 0.9 % NaCl or PBS (phosphate-buffered saline), fluids of physiologic significance. The introduction of more ingredients into the nanoparticulate formulation also increased

Table 6. Example of stability matrix for in 0.9 % NaCl (saline) or PBS solution

	0.05 % Calcium chloride/1 % F-68	0.05 % PMCQ/ 1 % F-68	0.05 % Calcium chloride/ 0.05 % PMCQ/1 % F-68
0.05 % HV Alginate	Unstable	Unstable	Unstable
0.05 % CS	Unstable	Unstable	Unstable
0.05 % HV/0.05 % CS	Unstable	Stable	Stable

Stability is defined as an inability to observe a Tyndall effect within few seconds following the addition of one component into another. The ratio C/S = 2/20 (mL/mL).

Hydrogel-Based Colloidal Polymeric System for Protein and Drug Delivery

Table 7. Example of stability matrix for in 0.9 % NaCl (saline) or PBS solution

	0.05 % Chitosan/ 1 % F-68	0.05 % Calcium chloride/1 % F-68	0.05 % Calcium chloride/ 0.05 % Chitosan/1 % F-68
0.05 % 3PP	Unstable	Unstable	Unstable
0.05 % X-52	Unstable	Unstable	Unstable
0.05 % 3PP/0.05 % X-52	Unstable	Unstable	Stable

Stability is defined as an inability to observe a Tyndall effect within few seconds following the addition of one component into another. The ratio C/S = 2/20 (mL/mL).

Table 8. Example of stability matrix for in 0.9 % NaCl (saline) or PBS solution

	0.05 % PMCG/ 1 % F-68	0.05 % Calcium chloride/1 % F-68	0.05 % Calcium chloride/ 0.05 % PMCG/1 % F-68
0.05 % Pectin	Unstable	Unstable	Unstable
0.05 % CS	Unstable	Unstable	Unstable
0.05 % Pectin/0.05 % CS	Stable	Unstable	Stable

Stability is defined as an inability to observe a Tyndall effect within few seconds following the addition of one component into another. The ratio C/S = 2/20 (mL/mL).

Table 9. Example of stability matrix for in 0.9 % NaCl (saline) or PBS solution

	0.05 % PMCG/ 1 % F-68	0.05 % Calcium chloride/1 % F-68	0.05 % Calcium chloride/ 0.05 % PMCG/1 % F-68
0.05 % Gellan	Unstable	Unstable	Unstable
0.05 % HV alginate	Coagulate	Unstable	Coagulate
0.05 % Gellan/0.05 % CS	Stable	Unstable	Stable

Stability is defined as an inability to observe a Tyndall effect within few seconds following the addition of one component into another. The ratio C/S = 2/20 (mL/mL).

their versatility in terms of association and delivery options [25, 26]. Tables 6 and 8–9 demonstrate that for some chemistries, certain ternary interactions also lead to a stable product.

For single-component systems, Krone et al. [24] demonstrated instabilities in buffered media. Bodmeier et al. [27] reported on instability of CT/TPP nanoparticles in 0.1 N HCl, while Calvo et al. [15] stated that their CT/TPP nanoparticles were stable in water, swelled when exposed to an acidic medium, but dissolved in a few minutes in 0.1 N HCl. Likewise, Bodmeier et al. [27] reported on calcium alginate nanoparticles stability in 0.1 N HCl, but a rapid disintegration in simulated intestinal fluids (USP XXI). Rajaonarivony et al. [28] did not report on the stability of their calcium alginate nanoparticles in physiologic solutions.

Calvo et al. [14, 15] used 5 % trehalose for stabilizing the particles. Likewise, their release data were obtained for a solution containing 5 % trehalose. We suspect that this non-metabolizable sugar was used because it imparts stability on particles, particularly in terms of preventing their swelling [29].

3.7
Freeze-Drying of Product

Fernandez-Urrusuno et al. [30] reported on development of a freeze-dried formulation of CT/TPP nanoparticles loaded with insulin. They tested various cryoprotective agents for their ability to maintain the original size/charge of particles upon reconstitution. The best medium appeared to be 5 % trehalose or sucrose. This was confirmed with unloaded nanoparticles from System #2 (Fig. 16). The original suspension, without the trehalose addition, exhibited a size of 185 nm (SD 25 nm).

3.8
Permeability Data

Several model proteins were chosen for the nanoparticle release studies. We have entrapped proteins in the native and cross-linked forms (PDA). The release rates were lower in saline than in pH 8.3 biocarbonate buffer or a low pH buffer

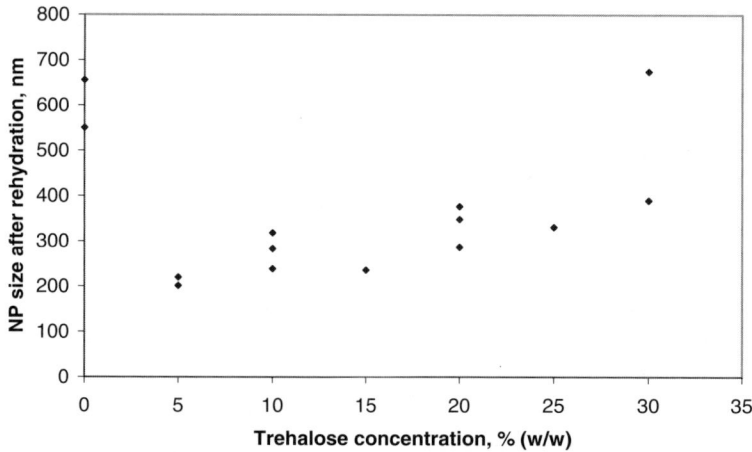

Fig. 16. Effect of trehalose on nanoparticle size (unloaded) after freeze-drying. The original size (without trehalose) was 195 nm (SD 3.5 nm). Nanoparticulate chemistry (System #2) was as in Fig. 5

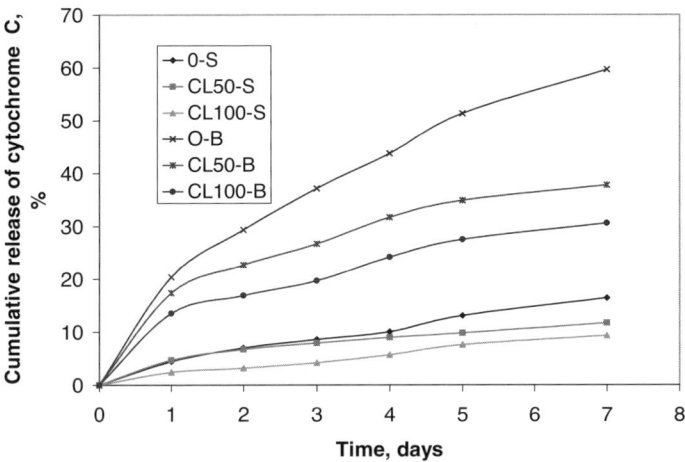

Fig. 17. Effect of cross-linking on cumulative release of cytochrome C (CC) vs time in different buffers. S – 0.9 % NaCl; B – bicarbonate pH 8.3. 0-CL – no cross-linking; 50-CL and 100-CK – cross-linked. Nanoparticulate chemistry was as follows – Core: 0.05 % HV, 0.05 % CS, 0.02 % CC and 0.05 % PDA for 50-CL or 0.01 % for 100-CL; Shell: 0.05 % SP, 0.075 % PMCG, 0.05 % $CaCl_2$, 1 % F-68. C/S ratio = 2/20 (mL/mL)

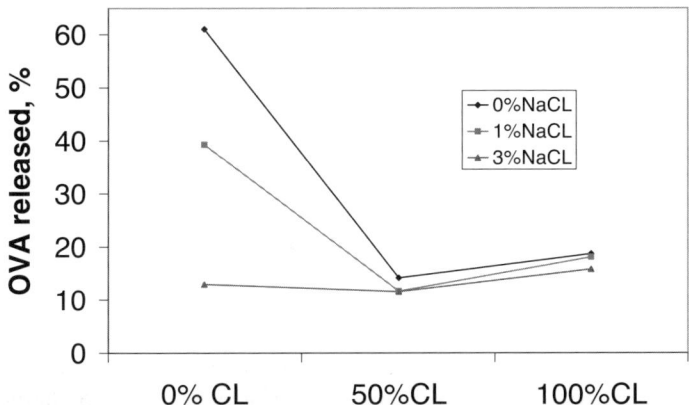

Fig. 18. Effect of salts and cross-linking on ovalbumin (OVA) release in 24 h. 0-CL (no cross-linking); 50-CL – and 100-CL – cross-linked. Nanoparticulate chemistry was same as in Fig. 3, except 0.05 % OVA was added in place of CC. PDA was 0.05 % for 50-CL, 0.1 % for 100-CL. C/S ratio = 2/20 (mL/mL)

(data not shown). Cross-linking progressively reduced release. The release data for cytochrome C (and for other proteins) support our belief that PDA can be used to reduce the rate of protein release (Fig. 17). To gain further insight into the mechanism of protein release, we investigated the effect of salts and temperature on protein release. Fig. 18 presents data for 0–3 % NaCl, showing a de-

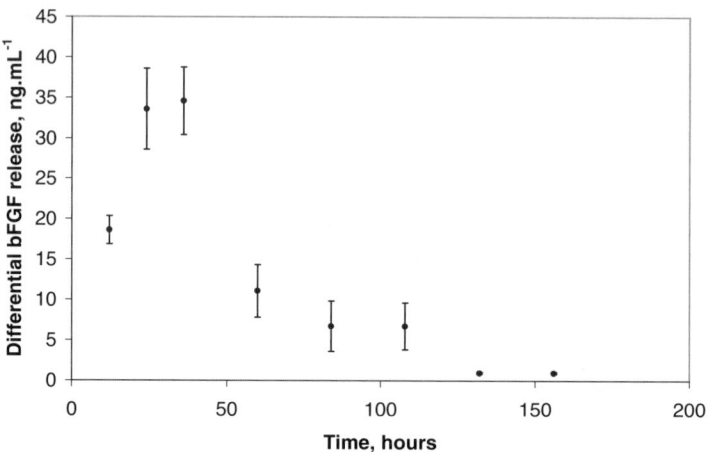

Fig. 19. Kinetics of *in vitro* bFGF release from nanoparticles, Nanoparticles were suspended in pH 7.5 buffer and buffer was totally replaced at different time points. Growth factor concentration was measured by a commercial ELISA. Nanoparticulate chemistry was same as in Fig. 9, except 0.00215 % BFGF was used in place of OVA. Concentration of nanoparticles was approximately 2.5 mg per batch (volume 0.5 mL). Data are presented as the mean ± SD ($n = 3$). C/S ratio = 4/20 (mL/mL)

crease of release rate with increased salt concentration. The release rate was not affected by temperature (37 °C vs 20 °C) with either non-cross-linked or cross-linked proteins (data not shown).

Another set of experiments involved bFGF (a basic protein) entrapment. The intent was to generate a suitable delivery vehicle applicable for wound healing and other areas where angiogenesis is of importance [31]. The biological activity of released bFGF was measured *in vitro* over period of 0.5–6.5 days by ELISA (Fig. 19). The proliferation test revealed a similar trend (data not shown). These data together indicated a high product activity within the nanoparticles, perhaps due to very gentle processing conditions. Release of radio-iodinated material in a second study permitted better accuracy and to carry the experiments over a longer period of time (Fig. 20). The release of bFGF was between 17 to 32 % over a 3-week period. This is a much slower release than expected for such a small molecule.

Fig. 21 shows data on small molecule (GMS) release. The rate of release was about 50 µg/day, representing about 5 %/day of the total entrapped amount. The amount of released GMS was based on a disc diffusion assay. A control experiment using an entrapped, non-conjugated GMS in reverse chemistry nanoparticles demonstrated a total absence of GMS in particles immediately after their preparation (data not shown).

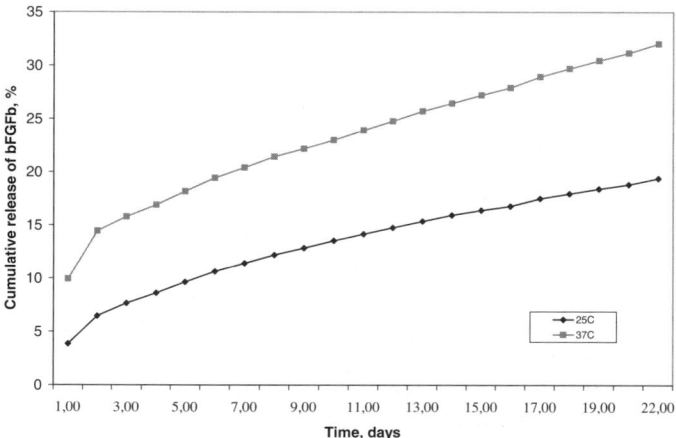

Fig. 20. *In vitro* cumulative release of radiolabeled fibroblast growth factor (bFGF) in PBS at 22 °C and 37 °C. Nanoparticle chemistry was same as in Fig. 3, except 0.00215 % bFGF was added in place of CC. The amount of bFGF captured (EE) was 7 %. This number also includes losses in the labeling process, as well. Concentration of nanoparticles was approximately 2.5 mg per batch (volume 0.5 mL)

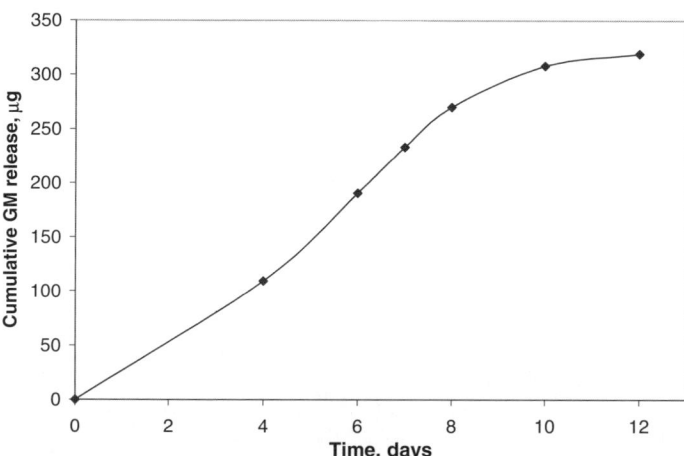

Fig. 21. Cumulative release of gentamycin sulfate conjugated to PEG in PBS buffer. Nanoparticulate chemistry was as follows – Core: 0.025 % HV, 0.025 % CS, 0.143 % GMS; Shell: 0.05 % SP, 0.065 % PMCG, 0.05 % $CaCl_2$, 1 % F-68. C/S ratio = 2/20 (mL/mL). Data are presented as the mean ± SD (not shown at this scale) ($n = 4$)

3.9
Vaccine Data

The product containing OVA could be considered as a prototype for delivery of antigens for vaccines [25]. Fig. 22 shows a dramatic difference in immunogenicity with nanoparticulate formulations of antigen (delivered SQ), compared to CRL-1005 adjuvant, used as a positive control, and two other controls, soluble antigen and antigen admixed to empty nanoparticles (System #2, Table 4). The nanoparticulate formulation produced a 10-fold higher titer. Another chemistry proved to give negative titres (System #1, Table 4). Fig. 23 demonstrates a usefulness of a PDA-cross-linked nanoparticulate formulation for oral immunization (see Table 4 for chemistry). Although the variations among the individual animals were substantial, the nanoparticulate formulation presented a clear benefit. The number of responders was also high. When no cross-linking was used and the product was applied orally, no immune response was observed (System #6, Table 4). Similar data were obtained for TT (System #3, Table 4). No cross-linked TT was applied orally, however. For cross-linked TT (SQ) data were similar to those presented in Fig. 22, although these data were not collected for non-cross-linked antigen (Fig. 24). No systematic studies of the effects of the nanoparticulate chemistry or dose were undertaken. A short communication on vaccine data was already published [32].

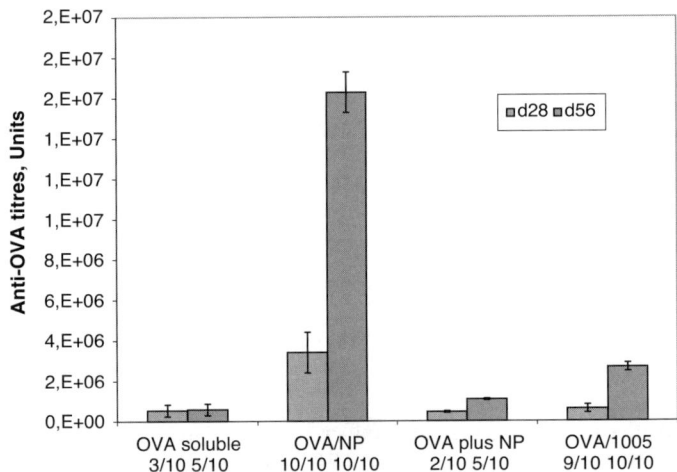

Fig. 22. Anti-OVA titres in mouse sera at day 28 and 56 for different experimental groups immunized subcutaneously. OVA soluble – injection of soluble antigen; OVA/NP – antigen formulated in nanoparticles; OVA plus NP – antigen admixed with empty nanoparticles; OVA/1005 – antigen formulated in CRL-1005 adjuvant. Data are presented as the mean ± SD. Number of responders given as a fraction. Nanoparticulate chemistry was as follows – Core: 0.025 % HV, 0.025 % GL, 0.0125 % HMP, 0.4 % OVA; Shell: 0.07 % PMCG, 0.05 % $CaCl_2$, 1 % F-68. C/S ratio = 2/20 (mL/mL). The antigen was not cross-linked

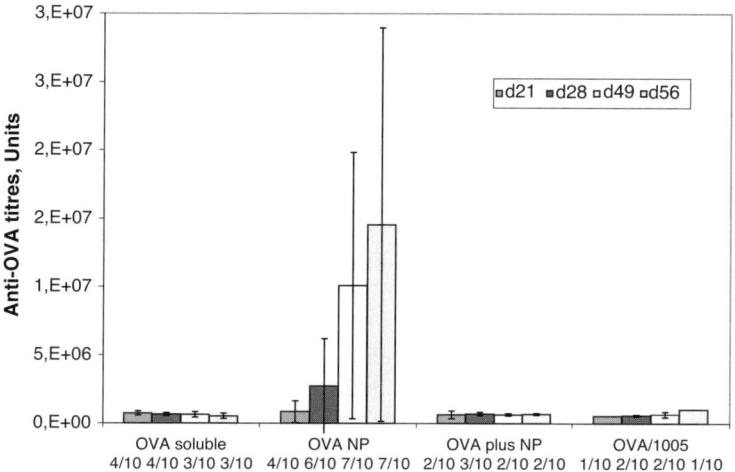

Fig. 23. Anti-OVA titres in mice at day 21, 28, 49 and 56 for different experimental groups immunized orally. Groups are the same as in Fig. 4. Data are presented as the mean ± SD. Number of responders as a fraction. Nanoparticulate chemistry was as follows – Core: 0.025 % HV, 0.025 % CS, 0.014 % PDA, 0.67 % OVA; Shell: 0.065 % PMCG, 0.05 % $CaCl_2$, 1 % F-68. C/S ratio = 2/20 (mL/mL). The antigen was cross-linked with PDA

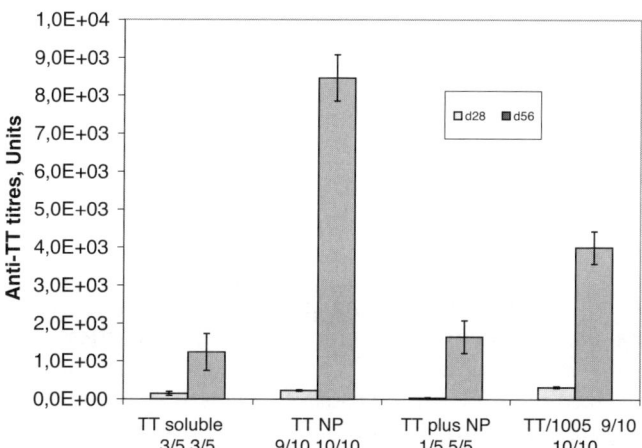

Fig. 24. Anti-TT titres in mice sera at day 28 and 56 for different experimental groups immunized subcutaneously. TT soluble – injection of soluble antigen; TT/NP – antigen formulated in nanoparticles; TT plus NP – antigen admixed with empty nanoparticles; TT/1005 – antigen formulated in CRL-1005 adjuvant. Data are mean±SD. Number of responders given as a fraction. Nanoparticulate chemistry was as follows – Core: 0.025 % HV, 0.05 % CS, 0.014 % PDA, 0.012 % TT; Shell: 0.05 % SP, 0.065 % PMCG, 0.05 % $CaCl_2$, 1 % F-68. C/S ratio =2/20 (mL/mL). The antigen was cross-linked with PDA

Calvo et al. [14] reported on antigen formulation and Alonso et al. [33] presented preliminary data on CT/TPP nanoparticles loaded with TT and DT antigens. Results are not available in a publication form; however, the data are remarkably similar to those presented above in this Section.

3.10
Biocompatibility Data

In terms of biocompatibility and inflammatory responses, the only acute effect noted was some respiratory difficulty in the animal receiving the highest dose. The only organ where we noticed an effect was the lung, in which the architecture was altered showing a dose-dependent effect. Low doses had no effect on lung morphology or the presence of inflammatory infiltrates. In all remaining organs, we did not detect any abnormality or see any evidence of inflammation even using the highest dose of nanoparticle suspension. The effects on the lung are likely to be due to a first pass effect as the nanoparticles reach the initial capillary bed with this route of administration.

4
Discussion

4.1
General

4.1.1
Scope

Over the last three decades, materials science has developed into an interdisciplinary field which encompasses the study of physical properties of matter and the means of manipulating those properties. From bulk materials to nanoparticles, the tools of materials science provide wide latitude of structural diversity and functional tunability in the development and design of new substances. Multicomposite polymers, smart polymers that respond to different physical or chemical stimuli and devices featuring an organization on a nanoscale molecular length are just few examples. Specific molecular properties can be produced in materials through the creation of structural hierarchies. The development of well-characterized methods for the controlled assembly of multicomponent nanostructures is highly desirable. Although it is clear that structures as complex as those found in the living world cannot yet be fabricated [34], our goal is to develop an assembly process for "simple" particles on the length scales of biological sub-cellular systems. Both the National Science Foundation [35] and the National Institute of Health [36] have identified nanotechnology/materials science as a critical technology area.

Much of the information available on intermolecular complexes formed between different classes of polymers (such as non-ionic polymers and polyelectrolytes) is qualitative in nature and heuristic because of the lack of suitable theory to interpret the experimental observations. The reasons for the formation of these self-assembled non-ionic morphologies are not well understood. In recent years, a number of theoretical predictions have been made about the phase behavior (phase separation) of block copolymers [37, 38]. Some theoretical constructs are also available for ionic interactions [39]. The latter interactions are the most versatile as they allow for the use of charged biopolymers such as DNA, polysaccharides and proteins (which are polyampholytes), as well as synthetic polyelectrolytes (such as methacrylates). Polyelectrolytes (as in the present work) permit the use of *water* as a solvent, which is *environmentally attractive* and has a major advantage for products that may be used as drug delivery systems in humans.

The electrostatic attraction between oppositely charged molecules is an adjustable driving force for structured material construction. The current synthetic routes of polymer production often offer many variations in size, topology, functionality and polydispersity. An electrostatically driven assembly of nanostructures allows for the controlled incorporation of materials available by synthetic routes. Biological macromolecules, nevertheless, offer superior polyfunctionality compared to synthetic macromolecules. We preferentially use them.

An ordered assembly, facilitated by charge, allows for a surface charge control of the nanoparticles. A high positive charge is considered to be a prerequisite for successful protein and gene delivery by means of nanoparticles. In addition, the degree of interaction of polymers (strong vs moderate) is important for subsequent release of the embedded protein within the cell interior. Such interactions are controllable by the rational selection of polymers and the underlying interactions. In addition to interactions between the constituent polymers, the incorporation of a charged biologic principle (e.g., protein) into the particle core introduces additional electrostatic interactions, particularly where the particle loading is high.

4.1.2
Uniqueness

Nanoparticles (NPs) are defined as objects smaller than 1 μm. In our work, NPs are generated using a multipolymeric water-based mixture with a minimum of two interacting pairs. The first polymer provides mechanical stability for the particles as well as a template (pre-formed structure) for a second interacting pair, while reacting almost instantaneously (a core gelling polysaccharide and a small inorganic ion); the second interaction develops rather slowly when the corona/shell polymers complex via a mutually interpenetrating (diffusion) process. At least one component of the corona polymers has a large molecular size. The corona-forming polymeric complex also provides some degree of permeability control. The use of a multicomponent system allows for a high degree of flexibility in choosing reacting pairs.

This work presents a novel water-based technology for producing nanoparticles. A superposition of several interacting principles to formulate delivery vehicles is also novel. Typically, one interacting pair of polymers does not result in a thermodynamically stable system [23]. Depending on the stoichiometry of components used, either soluble or insoluble complexes are formed. In addition, either the complex can undergo dissociation, or one member will exchange with another charged molecule that is present. A combination of several interacting principles may lead to more stable structures. Synthetic or natural polymers with repeating blocks can exhibit favorable interactions if the primary scale length is properly selected. Appropriate spacing leads to microdomain gelling and self-organization (assembly) of polymers into compact particles. Several physico-chemical principles can be applied to formulate suitable nanoparticle chemistries, e.g., charge neutralization by oppositely charged polymers; blending of flexible and rigid polymers; template-guided assembly; core/shell assembly; gelling vs precipitate formation; hydrophobic effects; charge transfer effects, and others. A combination of these approaches is a novel feature of our work.

Nanostructures primarily result from polyelectrolyte or interpolyelectrolyte complexes (PEC). The PEC (also referred to as symplex; [23]) is formed by the electrostatic interaction of oppositely charged polyelectrolytes (PE) in solution. The formation of PEC is governed by physical and chemical characteristics of the precursors, the environment where they react, and the technique used to introduce the reactants. Thus, the strength and location of ionic sites, polymer chain rigidity and precursor geometries, pH, temperature, solvent type, ionic strength, mixing intensity and other controllable factors will affect the PEC product. Three different types of PEC have been prepared in water [40]: (1) soluble PEC; (2) colloidal PEC systems, and; (3) two-phase systems of supernatant liquid and phase-separated PEC. These three systems are respectively characterized as:

- macroscopically homogeneous systems containing small PEC aggregates (micelles);
- macroscopically turbid systems with suspended PEC particles in the transition range to phase separation exhibiting an observable light scattering or Tyndall effect (TE), and;
- macroscopic systems with a second phase of easily isolable and washable, gelled particles (beads) or bulky precipitate (not suitable towards our goals).

The PECs of interest should be insoluble in water (and in other solvents) and thus would offer practical applications as microcapsules, nanoparticles, membranes and medical delivery vehicles. Soluble and colloidal PECs can be formed in a continuum of sizes, from a single two-molecular complex to insoluble symplex aggregates, depending on the ratio of PE solutions mixed together, on their relative molar mass, concentration and charge densities. Both kinds are of interest as delivery vehicles, the latter being termed nanoparticles (NPs), provided that their size is less than one micrometer (1–1000 nm). Two major principles underlie PEC formation: (1) the kinetic *diffusion process* of mutual entanglement between the two polymers, occurring at relatively short times, depending on the relative molar size differences, and; (2) the thermodynamic *rearrangement* of the already formed symplex aggregate due to conformational changes, disentanglement, etc., proceeding on longer time scales, a source of instability of PECs. The PEC rearrangement is a consequence of the phase separation, which occurs in an aqueous medium.

4.1.3
Nanoparticulate Production Technologies (Based on Polyelectrolyte Complexes)

Two routes are known for preparing NPs in the form of a macroscopic homogeneous colloidal system. According to Kabanov and Kabanov [23] such aggregates are obtained by reacting a high molar mass polyion species of a weak charge density with a much shorter (low molar mass) macromolecular counterion (oligomer) in a non-stoichiometric manner. We denote this case as a *dissimilar molar mass* PEC. Usually, the molar mixing ratio of reactants (precursors) of the high molar mass "host" polyion to the much smaller "guest" polycation in the reaction mixture is about 1:10 and at most 1:3. The complexes produced are typically soluble. The host system is denoted as a PE in solution in a receiving bath and the guest system as a PE solution used as a titrant. Through continued addition of the guest reactant, nanoparticles form as colloidal particles and will eventually aggregate and precipitate.

In the second route described by Dautzenberg et al. [41], NPs are generated via electrostatic interactions between oppositely charged PEs of *comparable molar mass* and of strong ionic groups at very low concentrations (in the range of 0.01–1 g/L). The resulting product maintains the non-stoichiometric ratio of reactants and is *insoluble*.

Both routes lead to the appearance of insoluble PECs and have a commonality of steps during the process of PE interaction. Route #1 involves the reaction between a large molar mass PE guest solution (polyanion) and a small molar mass PE host solution (oligocationic), the first one being slowly added to the second one. With a large excess of the host PE, a non-stoichiometric (soluble) complex with a high positive charge is formed initially. The size and mass of the soluble PEC particles grows with further addition of the guest PE. Over time, excess polycation chains incorporate into the complex. When the charge is neutralized, a stoichiometric complex precipitates. Since the guest polyanion is still being introduced, both types of PECs are present simultaneously. With continued addition of the anionic PE, the ratio of the stoichiometric complex to the non-stoichiometric complex increases; that is, the number of insoluble complexes increases relative to the number of soluble ones. (Further addition of the guest PE results in charge reversal on the NPs and the eventual dissolution of the particulate complexes) Thus, the product of practical interest is the insoluble stoichiometric complex unless the soluble non-stoichiometric one could be also used (in the area of drug delivery) as a covalent, soluble, polymer-drug adduct/conjugate.

Route #2 is similar to Route #1 in that a guest polyanion is metered into a host polycation solution. Cationic particles initially form and grow in size. As the charge of the complex is neutralized (i.e., at a stoichiometric charge ratio), aggregation occurs and a bulky precipitate falls out of solution. The main difference between Routes #1 and #2 is that *for Route #2 all products are insoluble under all reaction conditions* because of the comparable molar masses between host and guest species. Route #2 may prove to be more amenable for developing a functional drug delivery vehicle, but quasisoluble non-stoichiometric PEC obtained by Route #1 may be applied towards the same goal: the generation of insoluble and stable NPs. The size of the PEC appears to be the major factor and the issue of solubility or insolubility is a semantic one. Still another distinction is on thermodynamic grounds: the PEC formed from comparable molar mass components (Route #2) may not be thermodynamically stable as the entanglement and restructuring may take substantial time. In contrast, the PEC formed from the dissimilar molar size PE pair (Route #1) may reach a thermodynamically stable state under certain conditions (used in the present work). Thus, each route offers certain advantages (insolubility vs thermodynamic stability).

In the PEC system of interest, complexation effects are considered to be characterized by several interactions: cooperative, concerted, complementary and those due to microdomains [42]. Individual contributions are represented by a free energy thermodynamic function. For a PEC, the predominant term is the electrostatic interaction. Other terms include hydrogen bonding, hydrophobic interactions and van der Waals forces. Because individual components are difficult to evaluate [43] and their ratio is impossible to control independently, a superposition of different interactions is suggested. This approach is used in the present work.

4.2
Special

4.2.1
Novelty

The novelty of this technology is in the use of a multipolymeric water-soluble mixture of two interacting pairs enabling a template assembly of NPs. The technology features PEC production without the use of organic solvents and therefore the likelihood of a safe and patient-friendly product. The process offers high flexibility in choosing reacting pairs, a high degree of drug loading and a multiplicity of delivery events over a long period of time (sustained delivery). The superposition of several interacting principles to formulate delivery vehicles is a novel approach. Typically, one interacting pair of polymers does not result in a thermodynamically stable system. Depending on the stoichiometry of components used, either soluble or insoluble complexes are formed. In addition, either of the complexes can undergo dissociation or substitution if other charged molecules are present. A combination of several interacting principles can yield structures with specified characteristics and a plethora of nanoparticle formulations [44–46]. The shape of such particles is nearly spherical with a typical size range of a few hundred nanometers. An ordered assembly also allows for control of the NP surface charge. A moderate positive charge is considered to be a prerequisite for the successful uptake of NPs by cells.

The currently available nanotechnologies suffer from several drawbacks. PLGA (polylactide-glycolide co-polymer) is not suitable because of the use of a toxic organic solvent (methylene chloride), which impedes the regulatory approval process. Alkyl cyanoacrylate (ACA) nanotechnology, although now available as a water-based process, suffers from toxic breakdown products as recently demonstrated [4]. In contrast, the proposed nanotechnology is exclusively water-based and uses polymers, which are biocompatible and biodegradable and breakdown into products which are non-toxic and metabolizable into low molecular weight species. It has the potential to be developed into a successful vehicle for protein delivery.

4.2.2
Statistical Approach

Results of an extensive screening of individual reacting pairs provided us with a very broad basis for designing NPs [44–46]. Furthermore, our nanotechnology does not require a self-sustaining mechanical membrane system such as is necessary for stability of millimeter-sized microcapsules. Both the membrane-forming interactions and those leading to microscopic precipitates are conveniently used to generate useful NP carriers.

Because of the many possible interactions, not every attempt to design a particle is successful. Many technical hurdles exist. While it is essentially possible to arrive at a nanoparticle formation from almost any combination of charged polymers, ultimate success is achieved only when a satisfactory final product is obtained. This is perhaps due to the steps involved in separating the nanoparticles from the reaction mixture, which could lead to product instability and aggregation. It is not possible, at the present time, to predict *a priori* what chemical combination will be successful. Although we have already developed several vehicles capable of delivering several proteins, every entrapped system (drug or protein) will behave differently, particularly in terms of the stability of formulated nanoparticles. Thus, any reference to a success with a particular delivery vehicle will have to be considered with caution, as it has to be "reformulated" any time a new drug is considered.

In spite of the arbitrary aspects of design, two rational criteria for the selection of polymeric pairs used to assemble nanoparticles were employed in this work at. They are based on our extensive screen of polyelectrolytes resulting in strong interactions (capable of forming stable polymeric membranes) [44]. These criteria are: (1) the use of the combination of flexible and rigid polymeric chains, capable of mutual steric adjustments to form a complex; and (2) the use of small molecular weight polycations, facilitating their high diffusion within the polymeric network and good complex formation. Small molecular weight polycations (oligomers), such as protamine, spermine and PMCG were found to be very effective, polymeric complexing agents [44], even in our nanoparticulate formulations. The employment of these criteria substantially reduces our exploratory experiments for new nanoparticulate chemistries and enables us to expedite their development.

An additional criterion involves the use of the "engulfing" principle. The theoretical framework of Torza and Mason [47] and Pekarek et al. [48] describes an equilibrium configuration of two immiscible drops, readily predicted from the interfacial tensions of individual interaction terms. For a full engulfment of a drop 1 within a drop 2, the interfacial tension of phase 2, σ_{23} (vs air, 3) must be lower than that of phase 1, σ_{13} (vs air). In this case, phase 1 represents the core phase (in the specific case of Fig. 1, dispergated polyanionic liquid) and phase 2 the shell phase liquid (receiving bath anionic components). Although we did not intend to evaluate individual interfacial tension terms, there is a simple method to access the overall effect: a visual observation of introducing a droplet (e.g., by a Pasteur pipette) into a reservoir of phase 2 (either on the liquid top or via an injection inside the liquid). Other non-favorable polymer combinations ($\sigma_{23} > \sigma_{13}$) lead to a partial engulfment or non-engulfment state (a spreading of phase 2 over 1), without the required penetration of phase one into the phase two.

4.2.3
Surface Properties and Steric Stabilization

Surface properties, such as charge, porosity, size and the degree of hydrophilicity, are critical to the success of a particular nanoparticulate system. The employment of steric stabilizer, F-68, discouraged surface adhesion or interaction with various biological components. We used a simple entrapment of non-charged F-68 molecules within the shell layer of the nanoparticles during their assembly process. The presence of long chains of flexible polyethylene oxide chains (PEO) of the tri-block co-polymer (PEO/PPO/PEO; PPO is the polypropylene oxide moiety) at the nanoparticle surface created an entropic and an osmotic barrier to overcome protein or particle-particle interaction forces (van der Waals and hydrophobic), enhancing particle stability in suspensions [49]. The surface density of PEO and its effect on biodistribution for our nanoparticulate systems has not been optimized. Likewise, the effect of positive charge is not known (we cannot formulate a neutral control vehicle). Calvo et al. [14] reported on PEO and PEO/PPO entrapped nanoparticles. The amount of F-68 incorporated in our case was very low, based on our stoichiometric analysis. The surface density, however, seemed to be sufficient to provide adequate stability. The particle stability was assessed via particle compaction in a high-speed centrifuge (to recover the nanoparticles), while the particle-particle distance was pushed down close to the zeta potential distance, as determined by a simple test including different suspension media, relevant to physiologic applications (mild acidic or mild alkaline pH). We have not performed extensive tests to find a salt concentration which would induce flocculation.

We have demonstrated that both cationically and anionically charged particles can be formed. Antigen-loaded cationically charged particles could be beneficial for gut uptake, although the mechanism of the adjuvant effect is not currently known [50]. Cationically-modified antigen is taken up more vigorously compared with a non-modified form [51]. The enlarged size (due to a low pH in stomach) may still be suitable for their uptake by M-cells and macrophages. In addition to charge, nanoparticulate chemistry could contribute to the increased uptake of nanoparticles by mucus of the gastrointestinal tract. Chitosan-based particles could potentiate the paracellular route for absorption of particles, by binding to epithelia and opening the tight junctions [52]. The size of our product, particularly after swelling, probably does not allow for an endocytotic absorption pathway [53].

A second strategy aimed at modifying the surface properties of CT/PTT nanoparticles [54] has been the formation of PEG coating covalently linked to the surface of pre-formed nanoparticles. A carbodiimide reaction was employed to mediate a linkage (amide bond) between the free amine groups of the nanoparticles and methoxyPEG. This firmly attached PEG was found to decrease the positive surface charge significantly, improving their biocompatibility. This technique is more complicated than the current technology.

It should be noted that Calvo et al. [14, 15] demonstrated that no surface modifier is needed for suspension stability of their CT/TPP nanoparticles, free of PEO. However, based on our tests, such particles are not stable in saline or PBS buffer.

Data on freeze-drying demonstrated that trehalose could stabilize the final product and preserve the size. The ability of this sugar to protect biological structures from damage during desiccation is explained on the basis of its unique feature of reversibly absorbing water to produce the dihydrate from anhydrous polymorph crystalline form [55]. Preferential exclusion of the sugar from the surface of the macromolecule (particle) minimizes thermodynamic activity, which in turn preserves the preferred structure, by causing water molecules to pack more closely around the particle.

4.2.4
Drug Loading and Entrapment Efficiency

Drug loading (loading efficiency) capacity for water-soluble drugs is encouraged by creating an environment within the nanoparticles that is highly compatible with the solute. Since we achieved reasonable levels of drug loading, in some cases up to 50 %, we have not attempted a systematic optimization. In fact, the charged, entrapped drugs contribute towards the particle chemistry and become an integral part of the nanoparticles [25]. The entrapment efficiency reported for our OVA system was maximally 18 %, although not optimized

(Fig. 3). Calvo et al. [14] described CT-TPP nanoparticulate delivery vehicles for proteins and demonstrated high entrapment efficiencies (between 20 to 90 %). Their particles, however, dissolved at lower pH.

4.2.5
Mechanism of Release: Matrix vs Reservoir-Type Device

The new core-shell type of nanoparticle (assembled in a bottom-up process; [56]) might be considered as a matrix-type device, since the shell of the nanoparticle may not pose any additional resistance for the efflux of entrapped drugs. The drug is contained in the reservoir (core) from which it can exit by diffusion. If the diffusion of the drug is faster than the matrix degradation process, then the mechanism of drug release occurs mainly by diffusion. A rapid initial release, often observed with many delivery vehicles [57], is explained on the basis of the fraction of the drug that is adsorbed or weakly bound to the large surface area of the nanoparticles. The burst effect in our case was only moderate (see Figs. 12 and 14), a great advantage from the pharmacokinetic viewpoint. An exponential, delayed release rate (first-order kinetics) was probably due to the drug diffusion from the matrix. In case of a reservoir-type system (nanocapsules or particles with a shell), the drug release from nanoparticles would follow zero-order (linear) kinetics. The drug release from nanoparticles would occur by drug partitioning, the volume of the aqueous phase being the main factor. The partitioning assumes a high affinity of drug to the matrix. With a higher dilution (large sink), a faster release is usually observed. Our data indicated no dependency on sink volume (not reported here). The reservoir-type matrix is thus discounted.

4.2.6
Mechanism of Release: Release from Hydrogels

In the case of drug release from hydrogels, such release occurs mainly due to particle swelling, which can be controlled by formulation chemistry of the matrix (ionizable groups, degree of cross-linking) and by the environmental conditions (pH, temperature, ionic strength, etc.) [58]. Hydrogels that swell after contact with water permit diffusion of macromolecules throughout the entire matrix and their release from a porous structure. The size of the pores within the gel network is related to the extent of cross-linking, and the degree of swelling determines the protein release rate. By varying the degree of physical entanglement within the gel, by altering the number of cross-links between the polymer chains or by altering the interactions between the polymer matrix and molecules of interest, one can alter the release rate. Reducing the average free volume per molecule available to the solute retards the diffusive movement of a solute. However, the free volume model cannot adequately explain solute diffusivity in situations where the solute molecule is larger than the solvent molecule [59], which is our case. Swelling-controlled release is not controlled by Fickian diffusion, however [60]. This route is perhaps the decisive one in controlling protein release in our case. The physical entanglement is achieved via the degree of compactness of the (ionic) complex and by concentration of the cross-linker (PDA) for cross-linked nanoparticles.

4.2.7
Mechanism of Release:
Release from Hydrogels Possessing Ionizable and Pendant Groups – Swelling

Hydrogels possessing ionizable and frequent pendant groups exhibit increased swelling due to localization of charges within the hydrogel. The release of a drug from such a formulation involves the dissociation of the drug-polymer complex through the exchange of the drug with incoming counterions from the dissolution medium. Based on our data, we believe that the release mechanism of proteins from non-cross-linked nanoparticles is due to a combined effect of swelling, diffusion from the matrix associated complex and hydrophobic interactions. Ionic strength and pH influenced both swelling and ion exchange phenomena, effects observed in our case (Figs. 17–18). Remunan-Lopez and Bodmeier [61] reported on water uptake and swelling of two separate water-insoluble films generated from chitosan-TPP and alginate-calcium chloride as dependent on pH and concentration of the complexing agent (TPP or calcium chloride) as a method of controlling a release of encapsulated drugs. Calvo et al. [14] also implicated the swelling mechanism as the main factor contributing to the release of proteins from their CT/TPP nanoparticles, in addition to diffusion and dissociation from the polyelectrolyte complex. pH-dependent hydrolysis might be the crucial on the acidic side [62, 63].

4.2.8
Mechanism of Release: Hydrophobic Interactions

In addition to above phenomena, there is the sizable contribution of hydrophobic effects in our nanoparticulate system. As the opposite charges of interacting molecules are neutralized, and hydrophobicity rises, the particle is instantaneously formed. This is supported by our observations on the effect of ionic strength (Fig. 18), leading to an enhancement of release when the salt concentration is lowered. Hydrophobic interactions, in addition to ionic forces, were identified as an important mechanism for drug release from ion exchange resin [64], from the CT/TPP complex [62] and from theoretical calculations of forces involved in the assembly of polyelectrolytes [65].

4.2.9
A Need for Protein Cross-Linking: Schiff Base

We simulated the indefinitely large sink conditions, typical for *in vivo* application, by repeated and frequent replacement of the dissolution medium *in vitro*.

The release kinetics obtained from a buffer system may not have any clinical meaning, however. If extended release is required, therapeutic dose (nanoparticle loading) and release rates should be adjusted to lower rates. The protein release from PDA-cross-linked nanoparticles was retarded as expected. Our goal was the development of linkages that can be easily hydrolyzed. Instead of developing a permanent covalent bond under the reductive conditions such as in presence of cyanoborohydrate, non-reductive conditions were used in our studies, and this allowed us to take advantage of the relatively labile Schiff base. Aldehyde groups of PDA can react rapidly with primary amines, forming conjugated particles. The reaction chemistry is complicated and at later stages of the reaction multiple forms of complexes can be produced, followed by a possible rearrangement, leading to irreversible Amadori products [66]. The reversal of the Schiff base is possible by modulating pH and other conditions. This adduct readily dissociates at low pH, shifting the equilibrium by protonation. A full recovery of enzymatic activity is typically observed when protein is released from the Schiff base complex [67]. A release is also readily noted at higher pH, perhaps due to hydrolytic reactions (Fig. 17).

We observed no release after reductive amination of the Schiff base. The release of the residual amounts of protein after its cross-linking, which amounted to up to 70 % (Fig. 17), may only occur upon biodegradation of complexes *in vivo*. It is also possible that Amadori reactions could lead to a permanent, covalent immobilization of a part of the entrapped protein, only capable of release under matrix breakdown conditions.

4.2.10
Release of Small Size Proteins and Molecules

As CC is a small protein, its release from our cross-linked vehicles was moderately fast, a matter of 5–10 days. For radiolabeled bFGF, however, we observed a slow release, up to 3 weeks, with daily release amounting to few percent of the entrapped initial amount. The bFGF would be predicted to keep releasing beyond the tested time (data not collected).

As GMS is a small molecule it was not retained in any hydrogel formulation (data not presented). We increased the molecular mass and decreased the release rate by conjugating GMS to a large, polymeric species. This strategy proved correct. There is no simple strategy available for formulating and retaining small drug molecules via nanoparticulate delivery vehicles for extended periods of time.

4.2.11
Developing Safe Vaccine Vehicles: Bioavailability Issues

Data presented in this review also address a bioavailability issue, *in vivo* antigen delivery. Bioavailability measures how much of the drug molecule arrives at its site of action compared with how much is in the formulation delivered to the body. A significant limitation in the use of protein-based biologics is their short half-life in blood plasma, 1.5 to 150 min for some growth factors [68] or in other biological fluids. It is obvious that the use of naked protein *in vivo* (e.g., intravenously) has limited utility, requiring a need for more advanced delivery systems. The ideal method for local administration of a growth factor remains elusive, because of difficulties in satisfying many requirements. The material used in delivery should be non-toxic, not inhibiting wound healing, be compatible with the drug applied, preserving its activity, and delivering it with reproducible kinetics for several weeks. These criteria are even more important for substances that are quite labile and sensitive to insult (e.g., proteins, growth factors). The present study targets these needs. Similar arguments apply for intravenous and oral protein delivery. TT is known to be sensitive to denaturation [69]. Our water-based technology provides a gentle method for entrapping this antigen, opposite to solvent-based technologies [57]. At present, more data are being collected for biodistribution and bioavalability and will be reported in the near future.

In vivo antigen release from nanoparticles (at SQ and OR immunizations) was designed to take place for several days to weeks. However, rigorous experiments to show that no booster immunization is necessary have not been conducted. The most significant finding of the immunization study was the fact that only nanoparticulate PDA-cross-linked antigen was active in oral applications, in spite of relatively low PDA/OVA ratio used (0.021).

In terms of oral delivery, under normal conditions, only negligible amounts of proteins are absorbed intact through the gastrointestinal (GI) tract [70,71]. Ingested proteins are naturally fragmented into amino acids and short peptides by various enzymes present in the gut [72]. These proteolytic enzymes are found both in the lumen and in the mucosal epithelium; therefore, although proteins can be protected from the luminal enzymes by common encapsulated methods, they are still susceptible to degradation at the absorption site by various mucosal enzymes located in the intestinal walls.

Potentially successful encapsulated antigens must be able to transit stomach and upper part of the small intestine where the concentration of enzymes is highest and attach to ileum or the colon (where enzymatic activity is reduced) to increase the transit time in the gut. In the colon (and ileum), proteins are exposed not only to proteolysis by mucosal enzymes, but also to degradation by bacterial metabolism [71]. The co-administration of enzyme inhibitors (various trypsin and chymotrypsin inhibitors) is thus one strategy to reduce the degradation of protein antigens in the GI tract. Our technology can be easily adapted to allow for further potentiating the nanoparticulate delivery vehicle by means of including the enzyme inhibitors.

The second major obstacle of the oral delivery of proteins is the low permeability of proteins in the intestinal epithelium. The uptake of proteins is mediated by passive diffusion across the enterocytes (transcellular diffusion), paracellular diffusion (through intercellular spaces) and mostly by transcytosis (facilitated by receptor-mediated endocytosis). Erodible microcapsules and nanoparticles were shown to be absorbed intact through the GI tract and have opened the possibility for oral administration of proteins and antigens. The majority of available evidence suggests that the M-cells that overlie the lymphoid tissue of Peyer's patches (PP) of epithelium are the predominant site of uptake of microcapsules (1–5 µm) and nanoparticles via paracellular mechanism [73, 74] and to a low extent by transcytosis in case of nanoparticles [75, 76]. However, it should be noted that under normal conditions the absorption of particles is extremely low (<0.01 %). Improved particle uptake has been observed by using particles with mucoadhesive properties. It has been subsequently observed that mucoadhesive materials can additionally induce transcytosis (transient dilation of intercellular spaces) [77]. Thus the enhancement of particle uptake is higher than anticipated.

Many polymers we used to fabricate nanoparticles are mucoadhesive [78, 79]. Among them, e.g., alginate, carrageenans, pectin, etc., are typically used as the core polymers, thus hidden inside the nanoparticulate structure. The interior can be exposed while the nanoparticle undergoes transformation and break-up at the interaction with the gut epithelium. However, there is no doubt that the outer (shell) polymer Pluronic F-68, typically used to prevent aggregation, exhibits a very high degree of mucoadhesion, as do other members of the PEO family of polymers [79].

The current delivery strategies exhibit relatively low bioavailability, less than 10 % of the total protein dose in most cases [70]. Such bioavailability, however, may be sufficient in our case to initiate an immune response. Low bioavailabilities are a concern not only with regard to cost effectiveness, but also for dose reproducibility. Such uncertainty may increase data variability obtained in oral delivery of antigens. Nevertheless, nanoparticles may provide sufficient antigen protection from degradation in the gut lumen, in addition to increased bioavailability for prolonged periods of time directly at the surface of the mucous membrane. Encapsulated antigens are often more effective in the induction of a protective mucosal immune response compared to poorly immunogenic orally delivered soluble antigens [80]. Such a conclusion is in agreement with our results.

It might be concluded that the oral availability of proteins is largely determined by the ratio between the rates of adsorption and degradation.

4.2.12
Comparison with Other Water-Based Technologies

The technology described in this review provides a hydrophilic corona environment to enable entrapment of hydrophilic drugs and substances. Because the water-soluble core environment is highly compatible with the drug to be delivered, a high drug loading can be accomplished. The shell chemistry of opposite charge helps to retain the drug to some degree via polymeric complex formation. To achieve high loadings of products exhibiting an opposite charge at physiologic conditions (e.g., bFGF) a reverse chemistry system with the cationic core polymers is required. We will report on this system in an upcoming paper.

The hydrophilic delivery system described in this review can be extended to drugs with a low water-solubility (e.g., doxorubicin). Such compounds may be incorporated in CT/TPP nanoparticles by means of dextran sulfate complex prior to entrapment [54] or by dissolving them in a polar solvent (acetone, ethanol or acetonitrile) as demonstrated for the relatively hydrophobic peptide cyclosporin A [26, 81]. It is quite possible that this approach would work in a multicomponent polymer system as well.

Relatively weak covalent incorporation of drugs as applied in this work is advantageous. The current literature reveals that the covalent drug-carrier linkage does not facilitate success and often inhibits drug activity [82]. Typically, the term "cross-linking" means a formation of covalent bonds. In our case, the Schiff base complex is a rather weak covalent bond subjected to classical chemistry equilibration constraints (dissociation constant, shift in equilibrium due to reactant concentration and dissociation due to pH or temperature). The particles prepared with a Schiff base product complex seem to obey dissociation kinetics laws, with an increased release at higher temperatures and salt concentrations. As a consequence of the use of Schiff base cross-linkers, no harsh chemical conditions are applied, avoiding the use of organic solvents and tedious purification steps. The cross-linking agent (PDA) is a relatively non-immunogenic molecule [20] and becomes an integral part of the product.

Several formulations have been described as carriers for proteins and growth factors for a topical delivery, including methylcellulose, collagen/heparin sponge, fibrin scaffold, hyaluronic acid and PVA sponge, and Pluronic gel [83, 84]. The main drawback of these systems is that only a partial delay in the release of growth factors was observed because of the viscous nature of most of these vehicles. Another problem is due to a complicated production scheme of scaffolds and sponges and their excessive size. A controlled-release feature is absent in these formulations. Our approach is a simple process generating hydrogel-based vehicles distinguished by a delivery control feature [16].

Some potentially competitive technologies are discussed below. Among the methods using the preformed polymers one is of a mild chemical nature: gliadin nanoparticles [85]. A water/ethanol mixture is used for desolvation instead of organic solvents. If glutaradehyde cross-linking chemistry could be replaced by another less problematic approach (e.g., by the PDA chemistry we use), this technology could provide a very safe product. Among the water-based entrapment technologies, some have a potential to be scaled-down to a nanotechnology scale. Two synthetic biomimetics of silk and elastin proteins are available [86, 87]. Both polymers, initially water-soluble, can spontaneously form a gel via a thermally-induced (37 °C) or salt-induced process. Both offer high biocompatibility and controlled release (Fickian diffusion) features and could be adapted for small size particulate delivery. Another water-based nanotechnology is that of Katayose and Kataoka [88]. Their method is based on a self-assembly of hydrophobized polyelectrolytes in a dialysis system and as such is more suitable for hydrophobic drugs. Table 10 summarizes the present status of nanodelivery technologies suitable for hydrophilic drugs and lists some drawbacks.

Table 10. Some examples of hydrophilic nanoparticulate drug delivery systems

Composition	Method	Drug	Comment	Refs.
CT/TPP or	Gelation	None	Unstable in 0.1 N HCl,	[27]
Calcium alginate	Gelation	None	stable in 0.1 N HCl, but not in simulated intestinal fluids	
Calcium alginate and poly-L-lysine	Gelation	Doxorubicin, oligonucleotides		[28, 89]
CT	Emulsion coacervation	Gadolinium	Water-in-oil emulsion; high pH may be detrimental to some drugs	[90]
Sulfobutyl-PVA grafted to PLGA	Spontaneous assembly	Cytochrome C, TT	No surfactants, no solvents necessary	[91]
CT/TPP	Gelation	Proteins, TT, DT, insulin, oligonucleotides	Unstable in saline or PBS	[14, 15, 26, 30, 53]
Polyvinylpyrrolidone/ NIPAAM copolymer	Emulsion polymerization	None	n-Hexane used at processing	[92]
Human serum albumin	Desolvation, glutaraldehyde cross-linking or heat denaturation	None	Ethanol used for processing	[93]

4.2.13
Benefits of Controlled Delivery

Drug incorporation into delivery systems (polymeric matrices) offers many benefits, particularly via enhancing the therapeutic potential of many drugs. Among them are (1) *in vivo* predictability of release rate, with optimized peak plasma levels (pharmacokinetics) and thereby, reduced risk of adverse reactions; (2) reduced inconvenience of frequent dosing and, hence, improved patient compliance; (3) reduction of drugs systemic toxicity [94]; (4) drug stabilization and convenient pharmacodynamics [95]; and (5) effective delivery to a target site [96]. The main economic value could be in the re-formulation of drugs undergoing patent expiration. The problem, however, is the fact that the major forces in the pharmaceutical industry have not yet recognized drug delivery as an essential component for future innovation and growth [97]. The critical issues related to this question are how the delivery method impacts the manufacturing process, the integrity of the protein, drug bioavailability and the clinical toxicology [96]. We will address some of these issues in the forthcoming studies.

Acknowledgement. This work was in part supported by the National Institutes of Health grant 5R21 HL65982-02 and by the Apartment of Veterans Affairs. Data on antigen delivery in animal experiments were kindly provided in collaboration with Mark Newman of Vaxcel, Norcross, GA, at present with Epimmune, Inc., San Diego, CA. We also acknowledge Kenneth Hoffman of Rhodia, Nashville, for allowing us to use their laser diffraction sizing equipment and Xiang Gao for the use of his PCS/LDA instrument. Basic FGF was kindly provided by Dr. Judith Abraham of Scios, Inc. (presently with Chiron Corp., Emeryville, CA).

References

1. Buckley A, Davidson JM, Kamerath CD, Wolt TB, Woodward SC (1985) Sustained release of epidermal growth factor accelerates wound repair. Proc Natl Acad Sci USA 82:7340-7344
2. Vauthier-Holtzscherer C, Benabbou S, Spenlehauer G, Veillard M, Couvreur P (1991) Methodology for the preparation of ultra-dispersed polymer systems. STP Pharma Sci 2:109-116
3. Couvreur P, Roland M, Speiser P (1982) Biodegradable submicroscopic particles containing a biologically active substances and compositions containing them. US Patent 4,329,332
4. Cruz T, Gaspar R, Donato A, Lopes C (1997) Interaction between polyalkylcyanoacrylate nanoparticles and peritoneal macrophages: MTT metabolism, NBT reduction, and NO production. Pharm Res 14:73-79
5. Gallo JM, Hung CT, Perrier DG (1984) Analysis of albumin microsphere preparation. Int J Pharm 22:63-74
6. Marty JJ, Openheim RC, Speiser P (1978) Nanoparticles. A new colloidal drug delivery system. Pharm Acta Helv 53:17-23
7. Ibrahim H, Gurny R, Bindschaedler C, Doelker E, Buri P (1990) A new technology for preparation of drug monodispersed systems for controlled release to the eye. Proc Int Symp Contr Rel Bioact Mat, Controlled Release Society 17:303-304
8. Fessi H, Puisieux F, Devissaguet J-P, Ammoury N, Benita S (1989) Nanocapsule formation by interfacial polymer deposition following solvent displacement. Int J Pharm 55:R1-R4
9. Alonso MJ (1998) Nanoparticulate drug carrier technology. In: Cohen S, H. Bernstein H (eds). Microparticulate Systems for the Delivery of Proteins and Vaccines. New York: Marcel Dekker, pp 203-242
10. Allen C, Eisenberg A, Maysinger D (1999a) Copolymer drug carriers: Conjugates, micelles and microspheres. STP Pharma Sci 9:139-151
11. Allen C, Maysinger D, Eisenberg A (1999b) Nano-engineering block copolymer aggregates for drug delivery. Coll Surf Interfaces 16 B:3-27
12. Nakache E, Poulain N, Candau F, Orecchioni A-M, Irache JM (2000) Biopolymer and polymer nanoparticles and their biomedical application. In: Nalwa HS (ed), Handbook of Nanostructured Materials and Nanotechnology. Vol 5: Organics, Polymers, and Biological Materials. New York: Academic Press, pp 577-635
13. Fitch RM (1997) Polymer Colloids: A Comprehensive Introduction. San Diego: Academic Press, pp 81-98
14. Calvo P, Remunan-Lopez C, Vila-Jato JL, Alonso MJ (1997a) Novel hydrophilic chitosan-polyethylene oxide nanoparticles as protein carriers. J Appl Polymer Sci 63:125-132
15. Calvo P, Remunan-Lopez, Vila-Jato JL, Alonso MH (1997b) Chitosan and chitosan/ethylene oxide-propylene oxide block copolymer nanoparticles as novel carriers for proteins and vaccines. Pharmaceut Res 14:1431-1436

16. Prokop A (1999) Drug delivery system exhibiting permeability control. US patent pending; also PCT application WO9918934A1
17. Deacon S (1976) Assay of gentamicin in cerebrospinal fluid. J Clin Pathol 29:749–751
18. Prokop A, Holland CA, Kozlov E, Moore B, Tanner RD (2001) Water-based nanoparticulate polymeric system for protein delivery. Biotechnol Bioeng, 15:228–232.
19. Smith PK, Krohn RI, Hermanson GT, Mallia AK, Gartner FH, Provenzano MD, Fujimoto EK, Goeke NM, Olson BJ, Klenk DC (1985) Measurement of protein bicinchoninic acid [published erratum appears in Anal. Biochem 163:279 (1987)]. Anal Biochem 150:76–85
20. Fagnani R, Hagan MS, Bartholomew R (1990) Reduction of immunogenicity by covalent modification of murine and rabbit immunoglobulins with oxidized dextrans of low molecular weight. Cancer Res 50:3638–3645
21. Cheng H, Rudick MJ (1991) A membrane blotting method for following the time course of protein radioiodination background references using IODO-BEADS, Anal Biochem 198:191–193
22. Prokop A, Hunkeler DJ, DiMari S, Haralson MA, Wang TG (1998) Water soluble polymers for immunoisolation I: Complex coacervation and cytotoxicity. Advan Polymer Sci 136:1–52
23. Kabanov AV, Kabanov VA (1995) DNA complexes with polycations for the delivery of genetic material into cells. Bioconj Chem 6:7–20
24. Krone V, Magerstadt M, Walch A, Groner A, Hoffman D (1997) Pharmacological composition containing polyelectrolyte complexes in nanoparticulate form and at least one active agent. US Patent 5,700,459
25. Prokop A (1997) Micro-particulate and nano-particulate polymeric delivery system. US patent pending, 1997; also PCT application WO9918934A1
26. Janes KA, Calvo P, Alonso MJ (2001a) Polysaccharide colloidal particles as delivery systems for macromolecules. Advan Drug Delivery Revs 47:83–97
27. Bodmeier R, Chen H, Paeratakul O (1989) A novel approach to the oral delivery of micro- or nanoparticles. Pharm Res 6:413–417
28. Rajaonarivony M, Vauthier C, Couarraze G, Puisieux F, Couvreur P (1993) Development of a new drug carrier made from alginate. J Pharm Sci 82:912–917
29. Gribbonet EM, Hatley RHM, Gard T, Blair J, Kampinga J, Roser BJ (1996) Trehalose and novel hydrophobic sugar glasses in drug stabilization and delivery. In: Karsa DR, Stephenson RA (eds), Chemical Aspects of Drug Delivery Systems. Cambridge: The Royal Society of Chemistry, pp 138–145
30. Fernandez-Urrusuno R, Romani D, Calvo P, Vila-Jato JL, Alonso MJ (1999) Development of a freeze-dried formulation of insulin-loaded chitosan nanoparticles intended for nasal administration. S.T.P. Pharma Sci 9:429–436
31. Prokop A, Davidson JM, Dikov MM, Williams P (1999) Polymeric encapsulation system promoting angiogenesis. US patent pending; also PCT application WO0064954 A
32. Prokop A, Kozlov E, Newman GW, Newman MJ (2002) Water-based nanoparticulate polymeric system for protein delivery: Permeability control and vaccine application. Biotechnol Bioeng, accepted
33. Alonso MJ (1998) Nanoparticulate drug carrier technology. In: Cohen S, H. Bernstein H (eds), Microparticulate Systems for the Delivery of Proteins and Vaccines. New York: Marcel Dekker, pp 203–242
34. Decher G (1997) Fuzzy nanoassemblies: Towards layered polymeric multicomposites. Science 277:1232–1237
35. Anonymous (1995) NSF in a Changing World. The National Science Foundation's Strategic Plan. NSF 95-24. http://www.nsf.gov/nsf/nsfpub/straplan/contents.htm
36. Anonymous (1999) Bioengineering Nanotechnology Initiative.
37. Williams DRM, Fredrickson GH (1992) Cylindrical micelles in rigid-flexible diblock copolymers, Macromol 25:3561–3568

38. Holyst R, Schick M (1992) Correlations in the rigid-flexible diblock copolymer systems. J Chem Phys 96:730–739
39. Jayaram B, DiCapua FM, Beveridge DL (1991) A theoretical study of polyelectrolyte effects in protein-DNA interactions: Monte Carlo free energy simulations on the ion atmosphere contribution to the thermodynamics of lambda repressor-operator complex formation. J Am Chem Soc 113:5211–5221
40. Webster L, Huglin MB, Robb ID (1997) Complex formation between polyelectrolytes in dilute aqueous solution. Polymer 38:1373–1380
41. Dautzenberg H, Hartmann J, Grunewald S, Brand F (1996) Stoichiometry and structure of polyelectrolyte complex particles in diluted solutions. Ber Bunsenges Phys Chem 100:1024–1032
42. Tsuchida E, Abe K (1988) Polyelectrolyte complexes. In: Wilson AD, Prosser HJ (eds), Developments in Ionic Polymers – 2. London: Elsevier, pp 191–266
43. Chatterjee SK, Chhabra M, Rajabi FH, Farahani BV (1992) Thermodynamic studies of some copolymer-homopolymer-polyelectrolyte interactions. Polymer 33:3762–3766
44. Prokop A, Hunkeler DJ, DiMari S, Haralson MA, Wang TG (1998a) Water soluble polymers for immunoisolation I: Complex coacervation and cytotoxicity, Advan Polymer Sci 136:1–52
45. Prokop A, Hunkeler DJ, Wang TG (1998b) Water soluble polymers for immunoisolation II: Evaluation of multicomponent microencapsulation systems. Advan Polymer Sci 136:53–73
46. Wang T, Lacik I, Brissova M, Anilkumar AV, Prokop A, Hunkeler D, Green R, Shahrokhi K., Powers AC (1997) A new generation capsule and encapsulation system for immunoisolation of pancreatic islets. Nature Biotechnol 15:358–362
47. Torza S, Mason SG (1970) Three-phase interactions in shear and electrical fields. J Coll Interface Sci 33:67–83
48. Pekarek KJ, Jacob JS, Mathiowitz E (1994) Double-walled polymer microspheres for controlled drug release. Nature 367:158–260
49. Lee JH, Lee HB, Andrade JD (1995) Blood compatibility of polyethylene oxide surfaces. Progr Polymer Sci 20:1043–1079
50. Singh M, Brioned M, Ott G, O'Hagan D (2000) Cationic microparticles: A potent delivery system for DNA vaccines. Proc Natl Acad Sci USA 97:811–816
51. Altmann KG (1993) Effect of cationization on anti-hapten antibody response in sheep and mice. Immunol Cell Biology 71:517–525
52. Schipper NG, Olsson S, Hoogstraate JA, de Boer AG, Varum KM, Artursson P (1997) Chitosans as absorption enhancers for poorly adsorbable drugs 2: Mechanism of absorption enhancement. Pharm Res 14:923–929
53. Jung T, Kamm W, Breitenbach A, Kaiserling E, Ziao JX, Kissel T (2000) Biodegradable nanoparticles for oral delivery of peptides: Is there a role for polymers to affect mucosal uptake? Europ J Pharmac Biopharmaceut 50:147–160
54. Janes KA, Fresneau MP, Marazuela A, Fabra A, Alonso MJ (2001) Chitosan nanoparticles as delivery systems for doxorubicin. J Control Rel 73:255–267
55. Sussich F, Skopec C, Brady J, Cesaro A (2001) Reversible dehydration of trehalose and anhydrobiosis: from solution state to an exotic crystal? Carbohydrate Res 334:165–176
56. Shimomura M, Sawadaishi T (2001) Bottom-up strategy of materials fabrication: A new trend in nanotechnology of soft materials. Curr Opinion Coll Interf Sci 6:11–16
57. Singh M, Shirley B, Bajwa K, Samra E, Hora M, O'Hagan D (2001) Controlled release of recombinant insulin-like growth factor from a novel formulation of polylactide-co-glycolide microparticles. J Control Release 70:21–28
58. Peppas NA, Huang Y, Torres-Lugo M, Ward JH, Zhang J (2000) Physicochemical foundations and structural design of hydrogels in medicine and biology. Annu Rev Biomed Eng 2:9–16
59. Wesselingh JA, Bollen AM (1997) Multicomponent diffusivities from the free volume theory. Chem Eng Res Des 75(A6):590–602

60. Kim C-J, Nujoma YE (1995) Drug release from an erodible drug-polyelectrolyte complex. Europ Polymer J 31:937–940
61. Remunan-Lopez C, Bodmeier R (1997) Mechanical, water uptake and permeability properties of crosslinked chitosan glutamate and alginate films. J Control Rel 44:215–225
62. Yao K, Peng T, Xu M, Yuan C, Goosen MFA, Zhang Q, Ren L (1994) pH-dependent hydrolysis and drug release of chitosan/polyether interpenetrating polymer network hydrogel. Polymer Int. 34:213–219
63. Janes KA, Calvo P, Alonso MJ (2001) Polysaccharide colloidal particles as delivery systems for macromolecules. Advan Drug Delivery Revs 47:83–97
64. Kotov NA (1999) Layer-by-layer self-assembly: The contribution of hydrophobic interactions. NanoStruct Mat 12:789–796
65. Jaskari T, Vuorio M, Kontturi K, Manzanares JA, Hirvonen J (2001) Ion-exchange fibers and drugs: An equilibrium study. J Control Release 70:219–229
66. Ge S-J, Lee T-C (1997) Kinetic significance of the Schiff base reversion in the early-stage Maillard reaction of a phenylalanine-glucose aqueous model system. J Agric Food Chem 45:1619–1623
67. Chen S-S, Engel PC (1975) The equilibrium position of the reaction of bovine liver glutamate dehydrogenase with pyridoxal 5'-phosphate. Biochem J 147:351–358
68. Baldwin SP, Saltzman WM (1998) Materials for protein delivery in tissue engineering. Advan Drug Del Revs 33:71–86
69. Tobio M, Schwendeman SP, Guo Y, McIver J, Langer R, Alonso MJ (2000) Improved immunogenicity of a core-coated tetanus toxoid delivery system. Vaccine 18:618–622
70. Gomez-Orellana I, Paton DR (1998) Advances in the oral delivery of proteins. Exp Opinion Ther Patents 8:223–234
71. Matthews DM (1991) Protein Absorption. Wiley, New York
72. Woodley JF (1994) Enzymatic barriers for GI peptide and protein delivery. Crit Rev Ther Drug Carrier Syst 11:61–95
73. Jani P, Halbert GN, Landridgge J, Florence AT (1990) Nanoparticle uptake by the rat gastrointestinal mucose: Quantitation and particle size dependency. J Pharm Pharmacol 42:821–826
74. Damgge G, Aprahamian M, Marchais H, Benoit JP, Pinget M (1996) Intestinal absorption of PLGA microspheres in the gut. J Anat 189:491–501
75. Florence AT (1997) The oral absorption of micro- and nanoparticles: Neither exceptional nor unusual. Pharm Res 14:259–266
76. Puchel G, Montisci M-J, Dembri A, Durrer C, Duchene D (1997) Mucoadhesion of colloidal particulate system in the gastro-intestinal tract,. Europ J Pharm Biopharmaceut 44:25–31
77. Lehr C-M (1994) Bioadhesion technologies for the delivery of peptide and protein drugs to the gastrointestinal tract, Crit Rev Drug Carrier Syst 11:119–160
78. Hunt G, Pearney P, Kellaway I (1987) Mucoadhesive polymers in drug delivery systems, In: Johnson P. Lloyd-Jones JG (eds), Drug Delivery Systems. Fundamentals and Techniques. Ellis Horwood, Chichester, pp 180–199
79. Capron I, Yvon M, Muller G (1996) In vitro gastric stability of carrageenan. Food Hydrocol 10:239–244
80. Lavelle EC, Charif S, Thomas NW, Holland J, Davis SS (1995) The importance of gastrointestinal uptake of particles in the design of oral delivery systems. Adv Drug Del Revs 18:5–22
81. De Campos AM, Sanchez A, Alonso MJ (2001) Chitosan nanoparticles: A new vehicle for the improvement of the delivery of drugs to ocular surface. Application to cyclosporin A. Int J Pharm 224:159–168
82. Kaneda Y, Yamamoto Y, Kamada H, Tsunoda S, Tsutsumi Y, Hirano T, Mayumi Y (1998) Antitumor activity of tumor necrosis factor alpha conjugated with divinyl ether and maleic acid anhydride copolymer on solid tumors in mice. Cancer Res 58:290–295

83. Pierce GF, Tarpley JE, Janagihara D, Mustoe TA, Fox GM, Thomason A (1992) Platelet-derived growth factor (BB homodimer), transforming growth factor-beta, and basic fibroblast growth factor in dermal wound healing. Neovessel and matrix formation and cessation of repair. Amer J Pathol 140:1375–1388
84. Puolakkainen PA, Twardzik DR, Rauchalis JE, Pankey SC, Reed MJ, Gombotz WR (1995) The enhancement in wound healing by transforming growth factor β1 (TGFβ1) depends on the topical delivery system. J Surg Res 58:321–329
85. Couvreur P, Dubernet C, Puisieux F (1995) Controlled drug-delivery with nanoparticles. Current possibilities and future trends. Europ J Pharm Biopharm 41:2–13
86. Capello J, Crissman JW, Crissman M, Ferrari FA, Textor G, Wallis O, Whitledge JR, Zhou X, Burman D, Aukerman L, Stedronsky ER (1998) In-situ self-assembling polymer gel systems for administration, delivery, and release of drugs. J Control Release 53:105–117
87. Leon EJ, Verma N, Zhang S, Lauffenburger DA, Kamm RD (1998) Mechanical properties of a self-assembling oligopeptide matrix. J Biomat Sci Polymer Edn 9:297–312
88. Katayose S, Kataoka K (1997) Water-soluble polyanion complex associates of DNA and poly(ethylene glycol)-poly(L-lysine) block copolymer. Bioconjugate Chem 8:702–707
89. Aynie IC, Vauthier C, Fattal E, Foulquier M, Couvreur P (1998) Alginate nanoparticles as a novel carrier for antisense oligonucleotide. In: Diederichs JE, Muler R (eds), Future Strategies of Drug Delivery with Particulate Systems. Stuttgart: Medipharm Scientific Publisher, pp 5–10
90. Tokumitsu H, Ichikawa H, Fukumori Y (1999) Chitosan-gadopenteic acid complex nanoparticles for gadolinium neutron-capture therapy of cancer: Preparation by novel emulsion-droplet coalescence technique and characterization. Pharm Res 16:1839–1835
91. Jung T, Breitenbach A, Kissel T (2000) Sulfobutylated poly(vinyl alcohol)-graft-poly(lactide-co-glycolide) facilitate the preparation of small negatively charged biodegradable nanospheres for protein delivery. J Control Rel 67:157–169
92. Gaur U, Sahoo SK, De TK, Ghosh P, Maitra A, Ghosh PK (2000) Biodistribution of fluoresceinated dextran using novel nanoparticles evading reticuloendothelial system. Int J Pharm 202:1–10
93. Weber C, Coester C, Kreuter J, Langer K (2000) Desolvation process and surface characterisation of protein nanoparticles. Int J Pharm 194:91–102
94. Zhang X, Burt HM, VonHoff D, Dexter D, Mangold G, Degen D, Oktaba AM, Hunter WL (1997) An investigation of the antitumor activity and biodistribution of polymeric micellar paclitaxel. Cancer Chemother Pharmacol 40:81–86
95. Yokoyama M, Fukushima S, Uehara R, Okamoto K, Kataoka A, Sakurai Y, Ikano T (1998) Characterization of physical entrapment and chemical conjugation of adriamycin in polymeric micelles and their design for in vivo delivery to a solid tumor. J Control Release 50:79–92
96. Cleland JL, Daugherty A, Mrsny R (2001) Emerging protein delivery methods. Curr Opinion Biotechnol 12:212–219
97. Breimer DD (1998) Future challenges for drug delivery research. Advan Drug Del Revs 33:265–268

Received: October 2001

Author Index Volumes 101–160

Author Index Volumes 1–100 see Volume 100

de, Abajo, J. and *de la Campa, J.G.*: Processable Aromatic Polyimides. Vol. 140, pp. 23-60.
Adolf, D. B. see Ediger, M. D.: Vol. 116, pp. 73-110.
Aharoni, S. M. and *Edwards, S. F.*: Rigid Polymer Networks. Vol. 118, pp. 1-231.
Albertsson, A.-C., Varma, I. K.: Aliphatic Polyesters: Synthesis, Properties and Applications. Vol. 157, pp. 99–138.
Albertsson, A.-C. see Edlund, U.: Vol. 157, pp. 53-98.
Albertsson, A.-C. see Söderqvist Lindblad, M.: Vol. 157, pp. 139–161.
Albertsson, A.-C. see Stridsberg, K. M.: Vol. 157, pp. 27–51.
Améduri, B., Boutevin, B. and *Gramain, P.*: Synthesis of Block Copolymers by Radical Polymerization and Telomerization. Vol. 127, pp. 87-142.
Améduri, B. and *Boutevin, B.*: Synthesis and Properties of Fluorinated Telechelic Monodispersed Compounds. Vol. 102, pp. 133-170.
Amselem, S. see Domb, A. J.: Vol. 107, pp. 93-142.
Andrady, A. L.: Wavelenght Sensitivity in Polymer Photodegradation. Vol. 128, pp. 47-94.
Andreis, M. and *Koenig, J. L.*: Application of Nitrogen-15 NMR to Polymers. Vol. 124, pp. 191-238.
Angiolini, L. see Carlini, C.: Vol. 123, pp. 127-214.
Anseth, K. S., Newman, S. M. and *Bowman, C. N.*: Polymeric Dental Composites: Properties and Reaction Behavior of Multimethacrylate Dental Restorations. Vol. 122, pp. 177-218.
Antonietti, M. see Cölfen, H.: Vol. 150, pp. 67-187.
Armitage, B. A. see O'Brien, D. F.: Vol. 126, pp. 53-58.
Arndt, M. see Kaminski, W.: Vol. 127, pp. 143-187.
Arnold Jr., F. E. and *Arnold, F. E.*: Rigid-Rod Polymers and Molecular Composites. Vol. 117, pp. 257-296.
Arora, M. see Kumar, M.N.V.R.: Vol. 160, pp. 45-118.
Arshady, R.: Polymer Synthesis via Activated Esters: A New Dimension of Creativity in Macromolecular Chemistry. Vol. 111, pp. 1-42.

Bahar, I., Erman, B. and *Monnerie, L.*: Effect of Molecular Structure on Local Chain Dynamics: Analytical Approaches and Computational Methods. Vol. 116, pp. 145-206.
Ballauff, M. see Dingenouts, N.: Vol. 144, pp. 1-48.
Baltá-Calleja, F. J., González Arche, A., Ezquerra, T. A., Santa Cruz, C., Batallón, F., Frick, B. and *López Cabarcos, E.*: Structure and Properties of Ferroelectric Copolymers of Poly(vinylidene) Fluoride. Vol. 108, pp. 1-48.
Barnes, M. D. see Otaigbe, J.U.: Vol. 154, pp. 1-86.
Barshtein, G. R. and *Sabsai, O. Y.*: Compositions with Mineralorganic Fillers. Vol. 101, pp.1-28.
Baschnagel, J., Binder, K., Doruker, P., Gusev, A. A., Hahn, O., Kremer, K., Mattice, W. L., Müller-Plathe, F., Murat, M., Paul, W., Santos, S., Sutter, U. W., Tries, V.: Bridging the Gap Between Atomistic and Coarse-Grained Models of Polymers: Status and Perspectives. Vol. 152, pp. 41-156.
Batallán, F. see Baltá-Calleja, F. J.: Vol. 108, pp. 1-48.

Batog, A. E., Pet'ko, I. P., Penczek, P.: Aliphatic-Cycloaliphatic Epoxy Compounds and Polymers. Vol. 144, pp. 49-114.
Barton, J. see Hunkeler, D.: Vol. 112, pp. 115-134.
Bell, C. L. and *Peppas, N. A.*: Biomedical Membranes from Hydrogels and Interpolymer Complexes. Vol. 122, pp. 125-176.
Bellon-Maurel, A. see Calmon-Decriaud, A.: Vol. 135, pp. 207-226.
Bennett, D. E. see O'Brien, D. F.: Vol. 126, pp. 53-84.
Berry, G.C.: Static and Dynamic Light Scattering on Moderately Concentraded Solutions: Isotropic Solutions of Flexible and Rodlike Chains and Nematic Solutions of Rodlike Chains. Vol. 114, pp. 233-290.
Bershtein, V. A. and *Ryzhov, V. A.*: Far Infrared Spectroscopy of Polymers. Vol. 114, pp. 43-122.
Bigg, D. M.: Thermal Conductivity of Heterophase Polymer Compositions. Vol. 119, pp. 1-30.
Binder, K.: Phase Transitions in Polymer Blends and Block Copolymer Melts: Some Recent Developments. Vol. 112, pp. 115-134.
Binder, K.: Phase Transitions of Polymer Blends and Block Copolymer Melts in Thin Films. Vol. 138, pp. 1-90.
Binder, K. see Baschnagel, J.: Vol. 152, pp. 41-156.
Bird, R. B. see Curtiss, C. F.: Vol. 125, pp. 1-102.
Biswas, M. and *Mukherjee, A.*: Synthesis and Evaluation of Metal-Containing Polymers. Vol. 115, pp. 89-124.
Biswas, M. and *Sinha Ray, S.*: Recent Progress in Synthesis and Evaluation of Polymer-Montmorillonite Nanocomposites. Vol. 155, pp. 167-221.
Bolze, J. see Dingenouts, N.: Vol. 144, pp. 1-48.
Bosshard, C.: see Gubler, U.: Vol. 158, pp. 123-190.
Boutevin, B. and *Robin, J. J.*: Synthesis and Properties of Fluorinated Diols. Vol. 102. pp. 105-132.
Boutevin, B. see Amédouri, B.: Vol. 102, pp. 133-170.
Boutevin, B. see Améduri, B.: Vol. 127, pp. 87-142.
Bowman, C. N. see Anseth, K. S.: Vol. 122, pp. 177-218.
Boyd, R. H.: Prediction of Polymer Crystal Structures and Properties. Vol. 116, pp. 1-26.
Briber, R. M. see Hedrick, J. L.: Vol. 141, pp. 1-44.
Bronnikov, S. V., Vettegren, V. I. and *Frenkel, S. Y.*: Kinetics of Deformation and Relaxation in Highly Oriented Polymers. Vol. 125, pp. 103-146.
Brown, H. R. see Creton, C.: Vol. 156, pp. 53-135.
Bruza, K. J. see Kirchhoff, R. A.: Vol. 117, pp. 1-66.
Budkowski, A.: Interfacial Phenomena in Thin Polymer Films: Phase Coexistence and Segregation. Vol. 148, pp. 1-112.
Burban, J. H. see Cussler, E. L.: Vol. 110, pp. 67-80.
Burchard, W.: Solution Properties of Branched Macromolecules. Vol. 143, pp. 113-194.

Calmon-Decriaud, A. Bellon-Maurel, V., Silvestre, F.: Standard Methods for Testing the Aerobic Biodegradation of Polymeric Materials. Vol 135, pp. 207-226.
Cameron, N. R. and *Sherrington, D. C.*: High Internal Phase Emulsions (HIPEs)-Structure, Properties and Use in Polymer Preparation. Vol. 126, pp. 163-214.
de la Campa, J. G. see de Abajo, , J.: Vol. 140, pp. 23-60.
Candau, F. see Hunkeler, D.: Vol. 112, pp. 115-134.
Canelas, D. A. and *DeSimone, J. M.*: Polymerizations in Liquid and Supercritical Carbon Dioxide. Vol. 133, pp. 103-140.
Canva, M., Stegeman, G. I.: Quadratic Parametric Interactions in Organic Waveguides. Vol. 158, pp. 87-121.
Capek, I.: Kinetics of the Free-Radical Emulsion Polymerization of Vinyl Chloride. Vol. 120, pp. 135-206.
Capek, I.: Radical Polymerization of Polyoxyethylene Macromonomers in Disperse Systems. Vol. 145, pp. 1-56.

Capek, I.: Radical Polymerization of Polyoxyethylene Macromonomers in Disperse Systems. Vol. 146, pp. 1-56.
Capek, I. and *Chern, C.-S.:* Radical Polymerization in Direct Mini-Emulsion Systems. Vol. 155, pp. 101-166.
Carlesso, G. see Prokop, A.: Vol. 160, pp. 119–174.
Carlini, C. and *Angiolini, L.:* Polymers as Free Radical Photoinitiators. Vol. 123, pp. 127-214.
Carter, K. R. see Hedrick, J. L.: Vol. 141, pp. 1-44.
Casas-Vazquez, J. see Jou, D.: Vol. 120, pp. 207-266.
Chandrasekhar, V.: Polymer Solid Electrolytes: Synthesis and Structure. Vol 135, pp. 139-206
Chang, J. Y. see Han, M. J.: Vol. 153, pp. 1-36.
Charleux, B., Faust R.: Synthesis of Branched Polymers by Cationic Polymerization. Vol. 142, pp. 1-70.
Chen, P. see Jaffe, M.: Vol. 117, pp. 297-328.
Chern, C.-S. see Capek, I.: Vol. 155, pp. 101-166.
Choe, E.-W. see Jaffe, M.: Vol. 117, pp. 297-328.
Chow, T. S.: Glassy State Relaxation and Deformation in Polymers. Vol. 103, pp. 149-190.
Chung, T.-S. see Jaffe, M.: Vol. 117, pp. 297-328.
Cölfen, H. and *Antonietti, M.:* Field-Flow Fractionation Techniques for Polymer and Colloid Analysis. Vol. 150, pp. 67-187.
Comanita, B. see Roovers, J.: Vol. 142, pp. 179-228.
Connell, J. W. see Hergenrother, P. M.: Vol. 117, pp. 67-110.
Creton, C., Kramer, E. J., Brown, H. R., Hui, C.-Y.: Adhesion and Fracture of Interfaces Between Immiscible Polymers: From the Molecular to the Continuum Scale. Vol. 156, pp. 53-135.
Criado-Sancho, M. see Jou, D.: Vol. 120, pp. 207-266.
Curro, J.G. see Schweizer, K.S.: Vol. 116, pp. 319-378.
Curtiss, C. F. and *Bird, R. B.:* Statistical Mechanics of Transport Phenomena: Polymeric Liquid Mixtures. Vol. 125, pp. 1-102.
Cussler, E. L., Wang, K. L. and *Burban, J. H.:* Hydrogels as Separation Agents. Vol. 110, pp. 67-80.

Dalton, L. Nonlinear Optical Polymeric Materials: From Chromophore Design to Commercial Applications. Vol. 158, pp. 1-86.
Davidson, J.M. see Prokop, A.: Vol. 160, pp.119–174.
DeSimone, J. M. see Canelas D. A.: Vol. 133, pp. 103-140.
DiMari, S. see Prokop, A.: Vol. 136, pp. 1-52.
Dimonie, M. V. see Hunkeler, D.: Vol. 112, pp. 115-134.
Dingenouts, N., Bolze, J., Pötschke, D., Ballauf, M.: Analysis of Polymer Latexes by Small-Angle X-Ray Scattering. Vol. 144, pp. 1-48.
Dodd, L. R. and *Theodorou, D. N.:* Atomistic Monte Carlo Simulation and Continuum Mean Field Theory of the Structure and Equation of State Properties of Alkane and Polymer Melts. Vol. 116, pp. 249-282.
Doelker, E.: Cellulose Derivatives. Vol. 107, pp. 199-266.
Dolden, J. G.: Calculation of a Mesogenic Index with Emphasis Upon LC-Polyimides. Vol. 141, pp. 189-245.
Domb, A. J., Amselem, S., Shah, J. and *Maniar, M.:* Polyanhydrides: Synthesis and Characterization. Vol.107, pp. 93-142.
Domb, A.J. see Kumar, M.N.V.R.: Vol. 160, pp. 45–118.
Doruker, P. see Baschnagel, J.: Vol. 152, pp. 41-156.
Dubois, P. see Mecerreyes, D.: Vol. 147, pp. 1-60.
Dubrovskii, S. A. see Kazanskii, K. S.: Vol. 104, pp. 97-134.
Dunkin, I. R. see Steinke, J.: Vol. 123, pp. 81-126.
Dunson, D. L. see McGrath, J. E.: Vol. 140, pp. 61-106.

Eastmond, G. C.: Poly(ε-caprolactone) Blends. Vol.149, pp. 59-223.
Economy, J. and *Goranov, K.:* Thermotropic Liquid Crystalline Polymers for High Performance Applications. Vol. 117, pp. 221-256.

Ediger, M. D. and *Adolf, D. B.*: Brownian Dynamics Simulations of Local Polymer Dynamics. Vol. 116, pp. 73-110.
Edlund, U. Albertsson, A.-C.: Degradable Polymer Microspheres for Controlled Drug Delivery. Vol. 157, pp. 53-98.
Edwards, S. F. see Aharoni, S. M.: Vol. 118, pp. 1-231.
Endo, T. see Yagci, Y.: Vol. 127, pp. 59-86.
Engelhardt, H. and *Grosche, O.*: Capillary Electrophoresis in Polymer Analysis. Vol. 150, pp. 189-217.
Erman, B. see Bahar, I.: Vol. 116, pp. 145-206.
Ewen, B, Richter, D.: Neutron Spin Echo Investigations on the Segmental Dynamics of Polymers in Melts, Networks and Solutions. Vol. 134, pp. 1-130.
Ezquerra, T. A. see Baltá-Calleja, F. J.: Vol. 108, pp. 1-48.

Faust, R. see Charleux, B: Vol. 142, pp. 1-70.
Fekete, E see Pukánszky, B: Vol. 139, pp. 109-154.
Fendler, J.H.: Membrane-Mimetic Approach to Advanced Materials. Vol. 113, pp. 1-209.
Fetters, L. J. see Xu, Z.: Vol. 120, pp. 1-50.
Förster, S. and *Schmidt, M.*: Polyelectrolytes in Solution. Vol. 120, pp. 51-134.
Freire, J. J.: Conformational Properties of Branched Polymers: Theory and Simulations. Vol. 143, pp. 35-112.
Frenkel, S. Y. see Bronnikov, S. V.: Vol. 125, pp. 103-146.
Frick, B. see Baltá-Calleja, F. J.: Vol. 108, pp. 1-48.
Fridman, M. L.: see Terent´eva, J. P.: Vol. 101, pp. 29-64.
Fukui, K. see Otaigbe, J. U.: Vol. 154, pp. 1-86.
Funke, W.: Microgels-Intramolecularly Crosslinked Macromolecules with a Globular Structure. Vol. 136, pp. 137-232.

Galina, H.: Mean-Field Kinetic Modeling of Polymerization: The Smoluchowski Coagulation Equation. Vol. 137, pp. 135-172.
Ganesh, K. see Kishore, K.: Vol. 121, pp. 81-122.
Gaw, K. O. and *Kakimoto, M.*: Polyimide-Epoxy Composites. Vol. 140, pp. 107-136.
Geckeler, K. E. see Rivas, B.: Vol. 102, pp. 171-188.
Geckeler, K. E.: Soluble Polymer Supports for Liquid-Phase Synthesis. Vol. 121, pp. 31-80.
Gehrke, S. H.: Synthesis, Equilibrium Swelling, Kinetics Permeability and Applications of Environmentally Responsive Gels. Vol. 110, pp. 81-144.
de Gennes, P.-G.: Flexible Polymers in Nanopores. Vol. 138, pp. 91-106.
Giannelis, E.P., Krishnamoorti, R., Manias, E.: Polymer-Silicate Nanocomposites: Model Systems for Confined Polymers and Polymer Brushes. Vol. 138, pp. 107-148.
Godovsky, D. Y.: Device Applications of Polymer-Nanocomposites. Vol. 153, pp. 163-205.
Godovsky, D. Y.: Electron Behavior and Magnetic Properties Polymer-Nanocomposites. Vol. 119, pp. 79-122.
González Arche, A. see Baltá-Calleja, F. J.: Vol. 108, pp. 1-48.
Goranov, K. see Economy, J.: Vol. 117, pp. 221-256.
Gramain, P. see Améduri, B.: Vol. 127, pp. 87-142.
Grest, G.S.: Normal and Shear Forces Between Polymer Brushes. Vol. 138, pp. 149-184.
Grigorescu, G, Kulicke, W.-M.: Prediction of Viscoelastic Properties and Shear Stability of Polymers in Solution. Vol. 152, p. 1-40.
Grosberg, A. and Nechaev, S.: Polymer Topology. Vol. 106, pp. 1-30.
Grosche, O. see Engelhardt, H.: Vol. 150, pp. 189-217.
Grubbs, R., Risse, W. and *Novac, B.*: The Development of Well-defined Catalysts for Ring-Opening Olefin Metathesis. Vol. 102, pp. 47-72.
Gubler, U., Bosshard, C.: Molecular Design for Third-Order Nonlinear Optics. Vol. 158, pp. 123-190.
van Gunsteren, W. F. see Gusev, A. A.: Vol. 116, pp. 207-248.
Gusev, A. A., Müller-Plathe, F., van Gunsteren, W. F. and *Suter, U. W.*: Dynamics of Small Molecules in Bulk Polymers. Vol. 116, pp. 207-248.

Gusev, A. A. see Baschnagel, J.: Vol. 152, pp. 41-156.
Guillot, J. see Hunkeler, D.: Vol. 112, pp. 115-134.
Guyot, A. and *Tauer, K.*: Reactive Surfactants in Emulsion Polymerization. Vol. 111, pp. 43-66.

Hadjichristidis, N., Pispas, S., Pitsikalis, M., Iatrou, H., Vlahos, C.: Asymmetric Star Polymers Synthesis and Properties. Vol. 142, pp. 71-128.
Hadjichristidis, N. see Xu, Z.: Vol. 120, pp. 1-50.
Hadjichristidis, N. see Pitsikalis, M.: Vol. 135, pp. 1-138.
Hahn, O. see Baschnagel, J.: Vol. 152, pp. 41-156.
Hakkarainen, M.: Aliphatic Polyesters: Abiotic and Biotic Degradation and Degradation Products. Vol. 157, pp. 1-26.
Hall, H. K. see Penelle, J.: Vol. 102, pp. 73-104.
Hamley, I. W.: Crystallization in Block Copolymers. Vol. 148, pp. 113-138.
Hammouda, B.: SANS from Homogeneous Polymer Mixtures: A Unified Overview. Vol. 106, pp. 87-134.
Han, M.J. and Chang, J.Y.: Polynucleotide Analogues. Vol. 153, pp. 1-36.
Harada, A.: Design and Construction of Supramolecular Architectures Consisting of Cyclodextrins and Polymers. Vol. 133, pp. 141-192.
Haralson, M. A. see Prokop, A.: Vol. 136, pp. 1-52.
Hassan, C.M. and Peppas, N.A.: Structure and Applications of Poly(vinyl alcohol) Hydrogels Produced by Conventional Crosslinking or by Freezing/Thawing Methods. Vol. 153, pp. 37-65.
Hawker, C. J. Dentritic and Hyperbranched Macromolecules – Precisely Controlled Macromolecular Architectures. Vol. 147, pp. 113-160.
Hawker, C. J. see Hedrick, J. L.: Vol. 141, pp. 1-44.
Hedrick, J. L., Carter, K. R., Labadie, J. W., Miller, R. D., Volksen, W., Hawker, C. J., Yoon, D. Y., Russell, T. P., McGrath, J. E., Briber, R. M.: Nanoporous Polyimides. Vol. 141, pp. 1-44.
Hedrick, J. L., Labadie, J. W., Volksen, W. and *Hilborn, J. G.*: Nanoscopically Engineered Polyimides. Vol. 147, pp. 61-112.
Hedrick, J. L. see Hergenrother, P. M.: Vol. 117, pp. 67-110.
Hedrick, J. L. see Kiefer, J.: Vol. 147, pp. 161-247.
Hedrick, J.L. see McGrath, J. E.: Vol. 140, pp. 61-106.
Heinrich, G. and Klüppel, M.: Recent Advances in the Theory of Filler Networking in Elastomers. Vol. 160, pp. 1-44.
Heller, J.: Poly (Ortho Esters). Vol. 107, pp. 41-92.
Hemielec, A. A. see Hunkeler, D.: Vol. 112, pp. 115-134.
Hergenrother, P. M., Connell, J. W., Labadie, J. W. and *Hedrick, J. L.*: Poly(arylene ether)s Containing Heterocyclic Units. Vol. 117, pp. 67-110.
Hernández-Barajas, J. see Wandrey, C.: Vol. 145, pp. 123-182.
Hervet, H. see Léger, L.: Vol. 138, pp. 185-226.
Hilborn, J. G. see Hedrick, J. L.: Vol. 147, pp. 61-112.
Hilborn, J. G. see Kiefer, J.: Vol. 147, pp. 161-247.
Hiramatsu, N. see Matsushige, M.: Vol. 125, pp. 147-186.
Hirasa, O. see Suzuki, M.: Vol. 110, pp. 241-262.
Hirotsu, S.: Coexistence of Phases and the Nature of First-Order Transition in Poly-N-isopropylacrylamide Gels. Vol. 110, pp. 1-26.
Höcker, H. see Klee, D.: Vol. 149, pp. 1-57.
Hornsby, P.: Rheology, Compoundind and Processing of Filled Thermoplastics. Vol. 139, pp. 155-216.
Hui, C.-Y. see Creton, C.: Vol. 156, pp. 53-135
Hult, A., Johansson, M., Malmström, E.: Hyperbranched Polymers. Vol. 143, pp. 1-34.
Hunkeler, D., Candau, F., Pichot, C., Hemielec, A. E., Xie, T. Y., Barton, J., Vaskova, V., Guillot, J., Dimonie, M. V., Reichert, K. H.: Heterophase Polymerization: A Physical and Kinetic Comparision and Categorization. Vol. 112, pp. 115-134.

Hunkeler, D. see Prokop, A.: Vol. 136, pp. 1-52; 53-74.
Hunkeler, D see Wandrey, C.: Vol. 145, pp. 123-182.

Iatrou, H. see Hadjichristidis, N.: Vol. 142, pp. 71-128.
Ichikawa, T. see Yoshida, H.: Vol. 105, pp. 3-36.
Ihara, E. see Yasuda, H.: Vol. 133, pp. 53-102.
Ikada, Y. see Uyama, Y.: Vol. 137, pp. 1-40.
Ilavsky, M.: Effect on Phase Transition on Swelling and Mechanical Behavior of Synthetic Hydrogels. Vol. 109, pp. 173-206.
Imai, Y.: Rapid Synthesis of Polyimides from Nylon-Salt Monomers. Vol. 140, pp. 1-23.
Inomata, H. see Saito, S.: Vol. 106, pp. 207-232.
Inoue, S. see Sugimoto, H.: Vol. 146, pp. 39-120.
Irie, M.: Stimuli-Responsive Poly(N-isopropylacrylamide), Photo- and Chemical-Induced Phase Transitions. Vol. 110, pp. 49-66.
Ise, N. see Matsuoka, H.: Vol. 114, pp. 187-232.
*Ito, K., Kawaguchi, S,:*Poly(macronomers), Homo- and Copolymerization. Vol. 142, pp. 129-178.
Ivanov, A. E. see Zubov, V. P.: Vol. 104, pp. 135-176.

Jacob, S. and Kennedy, J.: Synthesis, Characterization and Properties of OCTA-ARM Polyisobutylene-Based Star Polymers. Vol. 146, pp. 1-38.
Jaffe, M., Chen, P., Choe, E.-W., Chung, T.-S. and *Makhija, S.*: High Performance Polymer Blends. Vol. 117, pp. 297-328.
Jancar, J.: Structure-Property Relationships in Thermoplastic Matrices. Vol. 139, pp. 1-66.
Jerôme, R.: see Mecerreyes, D.: Vol. 147, pp. 1-60.
Jiang, M., Li, M., Xiang, M. and Zhou, H.: Interpolymer Complexation and Miscibility and Enhancement by Hydrogen Bonding. Vol. 146, pp. 121-194.
Jin, J.: see Shim, H.-K.: Vol. 158, pp. 191-241.
Jo, W. H. and Yang, J. S.: Molecular Simulation Approaches for Multiphase Polymer Systems. Vol. 156, pp. 1-52.
Johansson, M. see Hult, A.: Vol. 143, pp. 1-34.
Joos-Müller, B. see Funke, W.: Vol. 136, pp. 137-232.
Jou, D., Casas-Vazquez, J. and *Criado-Sancho, M.*: Thermodynamics of Polymer Solutions under Flow: Phase Separation and Polymer Degradation. Vol. 120, pp. 207-266.

Kaetsu, I.: Radiation Synthesis of Polymeric Materials for Biomedical and Biochemical Applications. Vol. 105, pp. 81-98.
Kaji, K. see Kanaya, T.: Vol. 154, pp. 87-141.
Kakimoto, M. see Gaw, K. O.: Vol. 140, pp. 107-136.
Kaminski, W. and *Arndt, M.*: Metallocenes for Polymer Catalysis. Vol. 127, pp. 143-187.
Kammer, H. W., Kressler, H. and *Kummerloewe, C.*: Phase Behavior of Polymer Blends - Effects of Thermodynamics and Rheology. Vol. 106, pp. 31-86.
Kanaya, T. and Kaji, K.: Dynamcis in the Glassy State and Near the Glass Transition of Amorphous Polymers as Studied by Neutron Scattering. Vol. 154, pp. 87-141.
Kandyrin, L. B. and *Kuleznev, V. N.*: The Dependence of Viscosity on the Composition of Concentrated Dispersions and the Free Volume Concept of Disperse Systems. Vol. 103, pp. 103-148.
Kaneko, M. see Ramaraj, R.: Vol. 123, pp. 215-242.
Kang, E. T., Neoh, K. G. and *Tan, K. L.*: X-Ray Photoelectron Spectroscopic Studies of Electroactive Polymers. Vol. 106, pp. 135-190.
Karlsson, S. see Söderqvist Lindblad, M.: Vol. 157, pp. 139-161.
Kato, K. see Uyama, Y.: Vol. 137, pp. 1-40.
Kawaguchi, S. see Ito, K.: Vol. 142, p 129-178.
Kazanskii, K. S. and *Dubrovskii, S. A.*: Chemistry and Physics of „Agricultural" Hydrogels. Vol. 104, pp. 97-134.
Kennedy, J. P. see Jacob, S.: Vol. 146, pp. 1-38.
Kennedy, J. P. see Majoros, I.: Vol. 112, pp. 1-113.

Khokhlov, A., Starodybtzev, S. and *Vasilevskaya, V.*: Conformational Transitions of Polymer Gels: Theory and Experiment. Vol. 109, pp. 121-172.
Kiefer, J., Hedrick J. L. and *Hiborn, J. G.*: Macroporous Thermosets by Chemically Induced Phase Separation. Vol. 147, pp. 161-247.
Kilian, H. G. and *Pieper, T.*: Packing of Chain Segments. A Method for Describing X-Ray Patterns of Crystalline, Liquid Crystalline and Non-Crystalline Polymers. Vol. 108, pp. 49-90.
Kim, J. see Quirk, R.P.: Vol. 153, pp. 67-162.
Kishore, K. and *Ganesh, K.*: Polymers Containing Disulfide, Tetrasulfide, Diselenide and Ditelluride Linkages in the Main Chain. Vol. 121, pp. 81-122.
Kitamaru, R.: Phase Structure of Polyethylene and Other Crystalline Polymers by Solid-State ^{13}C/MNR. Vol. 137, pp 41-102.
Klee, D. and *Höcker, H.*: Polymers for Biomedical Applications: Improvement of the Interface Compatibility. Vol. 149, pp. 1-57.
Klier, J. see Scranton, A. B.: Vol. 122, pp. 1-54.
Klüppel, M. see Heinrich, G.: Vol. 160, pp 1-44.
Kobayashi, S., Shoda, S. and *Uyama, H.*: Enzymatic Polymerization and Oligomerization. Vol. 121, pp. 1-30.
Köhler, W. and *Schäfer, R.*: Polymer Analysis by Thermal-Diffusion Forced Rayleigh Scattering. Vol. 151, pp. 1-59.
Koenig, J. L. see Andreis, M.: Vol. 124, pp. 191-238.
Koike, T.: Viscoelastic Behavior of Epoxy Resins Before Crosslinking. Vol. 148, pp. 139-188.
Kokufuta, E.: Novel Applications for Stimulus-Sensitive Polymer Gels in the Preparation of Functional Immobilized Biocatalysts. Vol. 110, pp. 157-178.
Konno, M. see Saito, S.: Vol. 109, pp. 207-232.
Kopecek, J. see Putnam, D.: Vol. 122, pp. 55-124.
Koßmehl, G. see Schopf, G.: Vol. 129, pp. 1-145.
Kozlov, E. see Prokop, A.: Vol. 160, pp. 119-174.
Kramer, E. J. see Creton, C.: Vol. 156, pp. 53-135.
Kremer, K. see Baschnagel, J.: Vol. 152, pp. 41-156.
Kressler, J. see Kammer, H. W.: Vol. 106, pp. 31-86.
Kricheldorf, H. R.: Liquid-Cristalline Polyimides. Vol. 141, pp. 83-188.
Krishnamoorti, R. see Giannelis, E.P.: Vol. 138, pp. 107-148.
Kirchhoff, R. A. and *Bruza, K. J.*: Polymers from Benzocyclobutenes. Vol. 117, pp. 1-66.
Kuchanov, S. I.: Modern Aspects of Quantitative Theory of Free-Radical Copolymerization. Vol. 103, pp. 1-102.
Kuchanov, S. I.: Principles of Quantitive Description of Chemical Structure of Synthetic Polymers. Vol. 152, p. 157-202.
Kudaibergennow, S.E.: Recent Advances in Studying of Synthetic Polyampholytes in Solutions. Vol. 144, pp. 115-198.
Kuleznev, V. N. see Kandyrin, L. B.: Vol. 103, pp. 103-148.
Kulichkhin, S. G. see Malkin, A. Y.: Vol. 101, pp. 217-258.
Kulicke, W.-M. see Grigorescu, G.: Vol. 152, p. 1-40.
Kumar, M.N.V.R., Kumar, N., Domb, A.J. and *Arora, M.*: Pharmaceutical Polymeric Controlled Drug Delivery Systems. Vol. 160, pp. 45-118.
Kumar, N. see Kumar M.N.V.R.: Vol. 160, pp. 45-118.
Kummerloewe, C. see Kammer, H. W.: Vol. 106, pp. 31-86.
Kuznetsova, N. P. see Samsonov, G. V.: Vol. 104, pp. 1-50.Labadie, J. W. see Hergenrother, P. M.: Vol. 117, pp. 67-110.

Labadie, J. W. see Hedrick, J. L.: Vol. 141, pp. 1-44.
Labadie, J. W. see Hedrick, J. L.: Vol. 147, pp. 61-112.
Lamparski, H. G. see O´Brien, D. F.: Vol. 126, pp. 53-84.
Laschewsky, A.: Molecular Concepts, Self-Organisation and Properties of Polysoaps. Vol. 124, pp. 1-86.
Laso, M. see Leontidis, E.: Vol. 116, pp. 283-318.

Lazár, M. and RychlΩ, *R.*: Oxidation of Hydrocarbon Polymers. Vol. 102, pp. 189-222.
Lechowicz, J. see Galina, H.: Vol. 137, pp. 135-172.
Léger, L., Raphaël, E., Hervet, H.: Surface-Anchored Polymer Chains: Their Role in Adhesion and Friction. Vol. 138, pp. 185-226.
Lenz, R. W.: Biodegradable Polymers. Vol. 107, pp. 1-40.
Leontidis, E., de Pablo, J. J., Laso, M. and *Suter, U. W.*: A Critical Evaluation of Novel Algorithms for the Off-Lattice Monte Carlo Simulation of Condensed Polymer Phases. Vol. 116, pp. 283-318.
Lee, B. see Quirk, R.P: Vol. 153, pp. 67-162.
Lee, Y. see Quirk, R.P: Vol. 153, pp. 67-162.
Lesec, J. see Viovy, J.-L.: Vol. 114, pp. 1-42.
Li, M. see Jiang, M.: Vol. 146, pp. 121-194.
Liang, G. L. see Sumpter, B. G.: Vol. 116, pp. 27-72.
Lienert, K.-W.: Poly(ester-imide)s for Industrial Use. Vol. 141, pp. 45-82.
Lin, J. and *Sherrington, D. C.*: Recent Developments in the Synthesis, Thermostability and Liquid Crystal Properties of Aromatic Polyamides. Vol. 111, pp. 177-220.
Liu, Y. see Söderqvist Lindblad, M.: Vol. 157, pp. 139-161
López Cabarcos, E. see Baltá-Calleja, F. J.: Vol. 108, pp. 1-48.

Majoros, I., Nagy, A. and *Kennedy, J. P.*: Conventional and Living Carbocationic Polymerizations United. I. A Comprehensive Model and New Diagnostic Method to Probe the Mechanism of Homopolymerizations. Vol. 112, pp. 1-113.
Makhija, S. see Jaffe, M.: Vol. 117, pp. 297-328.
Malmström, E. see Hult, A.: Vol. 143, pp. 1-34.
Malkin, A. Y. and *Kulichkhin, S. G.*: Rheokinetics of Curing. Vol. 101, pp. 217-258.
Maniar, M. see Domb, A. J.: Vol. 107, pp. 93-142.
Manias, E., see Giannelis, E.P.: Vol. 138, pp. 107-148.
Mashima, K., Nakayama, Y. and *Nakamura, A.*: Recent Trends in Polymerization of a-Olefins Catalyzed by Organometallic Complexes of Early Transition Metals. Vol. 133, pp. 1-52.
Mathew, D. see Reghunadhan Nair, C.P.: Vol. 155, pp. 1-99.
Matsumoto, A.: Free-Radical Crosslinking Polymerization and Copolymerization of Multivinyl Compounds. Vol. 123, pp. 41-80.
Matsumoto, A. see Otsu, T.: Vol. 136, pp. 75-138.
Matsuoka, H. and *Ise, N.*: Small-Angle and Ultra-Small Angle Scattering Study of the Ordered Structure in Polyelectrolyte Solutions and Colloidal Dispersions. Vol. 114, pp. 187-232.
Matsushige, K., Hiramatsu, N. and *Okabe, H.*: Ultrasonic Spectroscopy for Polymeric Materials. Vol. 125, pp. 147-186.
Mattice, W. L. see Rehahn, M.: Vol. 131/132, pp. 1-475.
Mattice, W. L. see Baschnagel, J.: Vol. 152, p. 41-156.
Mays, W. see Xu, Z.: Vol. 120, pp. 1-50.
Mays, J.W. see Pitsikalis, M.: Vol.135, pp. 1-138.
McGrath, J. E. see Hedrick, J. L.: Vol. 141, pp. 1-44.
McGrath, J. E., Dunson, D. L., Hedrick, J. L.: Synthesis and Characterization of Segmented Polyimide-Polyorganosiloxane Copolymers. Vol. 140, pp. 61-106.
McLeish, T.C.B., Milner, S. T.: Entangled Dynamics and Melt Flow of Branched Polymers. Vol. 143, pp. 195-256.
Mecerreyes, D., Dubois, P. and *Jerôme, R.*: Novel Macromolecular Architectures Based on Aliphatic Polyesters: Relevance of the „Coordination-Insertion" Ring-Opening Polymerization. Vol. 147, pp. 1 -60.
Mecham, S. J. see McGrath, J. E.: Vol. 140, pp. 61-106.
Mikos, A. G. see Thomson, R. C.: Vol. 122, pp. 245-274.
Milner, S. T. see McLeish, T. C. B.: Vol. 143, pp. 195-256.
Mison, P. and Sillion, B.: Thermosetting Oligomers Containing Maleimides and Nadiimides End-Groups. Vol. 140, pp. 137-180.
Miyasaka, K.: PVA-Iodine Complexes: Formation, Structure and Properties. Vol. 108. pp. 91-130.
Miller, R. D. see Hedrick, J. L.: Vol. 141, pp. 1-44.

Monnerie, L. see Bahar, I.: Vol. 116, pp. 145-206.
Morishima, Y.: Photoinduced Electron Transfer in Amphiphilic Polyelectrolyte Systems. Vol. 104, pp. 51-96.
Morton M. see Quirk, R.P: Vol. 153, pp. 67-162
Mours, M. see Winter, H. H.: Vol. 134, pp. 165-234.
Müllen, K. see Scherf, U.: Vol. 123, pp. 1-40.
Müller-Plathe, F. see Gusev, A. A.: Vol. 116, pp. 207-248.
Müller-Plathe, F. see Baschnagel, J.: Vol. 152, p. 41-156.
Mukerherjee, A. see Biswas, M.: Vol. 115, pp. 89-124.
Murat, M. see Baschnagel, J.: Vol. 152, p. 41-156.
Mylnikov, V.: Photoconducting Polymers. Vol. 115, pp. 1-88.

Nagy, A. see Majoros, I.: Vol. 112, pp. 1-11.
Nakamura, A. see Mashima, K.: Vol. 133, pp. 1-52.
Nakayama, Y. see Mashima, K.: Vol. 133, pp. 1-52.
Narasinham, B., Peppas, N. A.: The Physics of Polymer Dissolution: Modeling Approaches and Experimental Behavior. Vol. 128, pp. 157-208.
Nechaev, S. see Grosberg, A.: Vol. 106, pp. 1-30.
Neoh, K. G. see Kang, E. T.: Vol. 106, pp. 135-190.
Newman, S. M. see Anseth, K. S.: Vol. 122, pp. 177-218.
Nijenhuis, K. te: Thermoreversible Networks. Vol. 130, pp. 1-252.
Ninan, K.N. see Reghunadhan Nair, C. P.: Vol. 155, pp. 1-99.
Noid, D. W. see Otaigbe, J.U.: Vol. 154, pp. 1-86.
Noid, D. W. see Sumpter, B. G.: Vol. 116, pp. 27-72.
Novac, B. see Grubbs, R.: Vol. 102, pp. 47-72.
Novikov, V. V. see Privalko, V. P.: Vol. 119, pp. 31-78.

O'Brien, D. F., Armitage, B. A., Bennett, D. E. and *Lamparski, H. G.*: Polymerization and Domain Formation in Lipid Assemblies. Vol. 126, pp. 53-84.
Ogasawara, M.: Application of Pulse Radiolysis to the Study of Polymers and Polymerizations. Vol.105, pp. 37-80.
Okabe, H. see Matsushige, K.: Vol. 125, pp. 147-186.
Okada, M.: Ring-Opening Polymerization of Bicyclic and Spiro Compounds. Reactivities and Polymerization Mechanisms. Vol. 102, pp. 1-46.
Okano, T.: Molecular Design of Temperature-Responsive Polymers as Intelligent Materials. Vol. 110, pp. 179-198.
Okay, O. see Funke, W.: Vol. 136, pp. 137-232.
Onuki, A.: Theory of Phase Transition in Polymer Gels. Vol. 109, pp. 63-120.
Osad'ko, I.S.: Selective Spectroscopy of Chromophore Doped Polymers and Glasses. Vol. 114, pp. 123-186.
Otaigbe, J. U., Barnes, M. D., Fukui, K., Sumpter, B. G., Noid, D. W.: Generation, Characterization, and Modeling of Polymer Micro- and Nano-Particles. Vol. 154, pp. 1-86.
Otsu, T., Matsumoto, A.: Controlled Synthesis of Polymers Using the Iniferter Technique: Developments in Living Radical Polymerization. Vol. 136, pp. 75-138.

de Pablo, J. J. see Leontidis, E.: Vol. 116, pp. 283-318.
Padias, A. B. see Penelle, J.: Vol. 102, pp. 73-104.
Pascault, J.-P. see Williams, R. J. J.: Vol. 128, pp. 95-156.
Pasch, H.: Analysis of Complex Polymers by Interaction Chromatography. Vol. 128, pp. 1-46.
Pasch, H.: Hyphenated Techniques in Liquid Chromatography of Polymers. Vol. 150, pp. 1-66.
Paul, W. see Baschnagel, J.: Vol. 152, p. 41-156.
Penczek, P. see Batog, A. E.: Vol. 144, pp. 49-114.
Penelle, J., Hall, H. K., Padias, A. B. and *Tanaka, H.*: Captodative Olefins in Polymer Chemistry. Vol. 102, pp. 73-104.

Peppas, N. A. see Bell, C. L.: Vol. 122, pp. 125-176.
Peppas, N.A. see Hassan, C.M.: Vol. 153, pp. 37-65
Peppas, N. A. see Narasimhan, B.: Vol. 128, pp. 157-208.
Pet'ko, I. P. see Batog, A. E.: Vol. 144, pp. 49-114.
Pichot, C. see Hunkeler, D.: Vol. 112, pp. 115-134.
Pieper, T. see Kilian, H. G.: Vol. 108, pp. 49-90.
Pispas, S. see Pitsikalis, M.: Vol. 135, pp. 1-138.
Pispas, S. see Hadjichristidis: Vol. 142, pp. 71-128.
Pitsikalis, M., Pispas, S., Mays, J. W., Hadjichristidis, N.: Nonlinear Block Copolymer Architectures. Vol. 135, pp. 1-138.
Pitsikalis, M. see Hadjichristidis: Vol. 142, pp. 71-128.
Pötschke, D. see Dingenouts, N.: Vol 144, pp. 1-48.
Pokrovskii, V. N.: The Mesoscopic Theory of the Slow Relaxation of Linear Macromolecules. Vol. 154, pp. 143-219.
Pospíšil, J.: Functionalized Oligomers and Polymers as Stabilizers for Conventional Polymers. Vol. 101, pp. 65-168.
Pospíšil, J.: Aromatic and Heterocyclic Amines in Polymer Stabilization. Vol. 124, pp. 87-190.
Powers, A. C. see Prokop, A.: Vol. 136, pp. 53-74.
Priddy, D. B.: Recent Advances in Styrene Polymerization. Vol. 111, pp. 67-114.
Priddy, D. B.: Thermal Discoloration Chemistry of Styrene-co-Acrylonitrile. Vol. 121, pp. 123-154.
Privalko, V. P. and *Novikov, V. V.:* Model Treatments of the Heat Conductivity of Heterogeneous Polymers. Vol. 119, pp 31-78.
Prokop, A., Hunkeler, D., Powers, A. C., Whitesell, R. R., Wang, T. G.: Water Soluble Polymers for Immunoisolation II: Evaluation of Multicomponent Microencapsulation Systems. Vol. 136, pp. 53-74.
Prokop, A., Hunkeler, D., DiMari, S., Haralson, M. A., Wang, T. G.: Water Soluble Polymers for Immunoisolation I: Complex Coacervation and Cytotoxicity. Vol. 136, pp. 1-52.
Prokop, A., Kozlov, E., Carlesso, G. and Davidsen, J.M.: Hydrogel-Based Colloidal Polymeric System for Protein and Drug Delivery: Physical and Chemical Characterization, Permeability Control and Applications. Vol. 160, pp. 119-174.
Pukánszky, B. and *Fekete, E.:* Adhesion and Surface Modification. Vol. 139, pp. 109-154.
Putnam, D. and *Kopecek, J.:* Polymer Conjugates with Anticancer Acitivity. Vol. 122, pp. 55-124.

Quirk, R.P. and Yoo, T., Lee, Y., M., Kim, J. and Lee, B.: Applications of 1,1-Diphenylethylene Chemistry in Anionic Synthesis of Polymers with Controlled Structures. Vol. 153, pp. 67-162.

Ramaraj, R. and *Kaneko, M.:* Metal Complex in Polymer Membrane as a Model for Photosynthetic Oxygen Evolving Center. Vol. 123, pp. 215-242.
Rangarajan, B. see Scranton, A. B.: Vol. 122, pp. 1-54.
Ranucci, E. see Söderqvist Lindblad, M.: Vol. 157, pp. 139–161.
Raphaël, E. see Léger, L.: Vol. 138, pp. 185-226.
Reddinger, J. L. and *Reynolds, J. R.:* Molecular Engineering of π-Conjugated Polymers. Vol. 145, pp. 57-122.
Reghunadhan Nair, C.P., Mathew, D. and *Ninan, K.N.,* : Cyanate Ester Resins, Recent Developments. Vol. 155, pp. 1-99.
Reichert, K. H. see Hunkeler, D.: Vol. 112, pp. 115-134.
Rehahn, M., Mattice, W. L., Suter, U. W.: Rotational Isomeric State Models in Macromolecular Systems. Vol. 131/132, pp. 1-475.
Reynolds, J.R. see Reddinger, J. L.: Vol. 145, pp. 57-122.
Richter, D. see Ewen, B.: Vol. 134, pp.1-130.
Risse, W. see Grubbs, R.: Vol. 102, pp. 47-72.
Rivas, B. L. and *Geckeler, K. E.:* Synthesis and Metal Complexation of Poly(ethyleneimine) and Derivatives. Vol. 102, pp. 171-188.
Robin, J. J. see Boutevin, B.: Vol. 102, pp. 105-132.

Roe, R.-J.: MD Simulation Study of Glass Transition and Short Time Dynamics in Polymer Liquids. Vol. 116, pp. 111-114.
Roovers, J., Comanita, B.: Dendrimers and Dendrimer-Polymer Hybrids. Vol. 142, pp 179-228.
Rothon, R. N.: Mineral Fillers in Thermoplastics: Filler Manufacture and Characterisation. Vol. 139, pp. 67-108.
Rozenberg, B. A. see Williams, R. J. J.: Vol. 128, pp. 95-156.
Ruckenstein, E.: Concentrated Emulsion Polymerization. Vol. 127, pp. 1-58.
Rusanov, A. L.: Novel Bis (Naphtalic Anhydrides) and Their Polyheteroarylenes with Improved Processability. Vol. 111, pp. 115-176.
Russel, T. P. see Hedrick, J. L.: Vol. 141, pp. 1-44.
Rychlý, J. see Lazár, M.: Vol. 102, pp. 189-222.
Ryner, M. see Stridsberg, K. M.: Vol. 157, pp. 27–51.
Ryzhov, V. A. see Bershtein, V. A.: Vol. 114, pp. 43-122.

Sabsai, O. Y. see Barshtein, G. R.: Vol. 101, pp. 1-28.
Saburov, V. V. see Zubov, V. P.: Vol. 104, pp. 135-176.
Saito, S., Konno, M. and *Inomata, H.*: Volume Phase Transition of N-Alkylacrylamide Gels. Vol. 109, pp. 207-232.
Samsonov, G. V. and *Kuznetsova, N. P.*: Crosslinked Polyelectrolytes in Biology. Vol. 104, pp. 1-50.
Santa Cruz, C. see Baltá-Calleja, F. J.: Vol. 108, pp. 1-48.
Santos, S. see Baschnagel, J.: Vol. 152, p. 41-156.
Sato, T. and *Teramoto, A.*: Concentrated Solutions of Liquid-Christalline Polymers. Vol. 126, pp. 85-162.
Schäfer R. see Köhler, W.: Vol. 151, pp. 1-59.
Scherf, U. and *Müllen, K.*: The Synthesis of Ladder Polymers. Vol. 123, pp. 1-40.
Schmidt, M. see Förster, S.: Vol. 120, pp. 51-134.
Schopf, G. and *Koßmehl, G.*: Polythiophenes - Electrically Conductive Polymers. Vol. 129, pp. 1-145.
Schweizer, K. S.: Prism Theory of the Structure, Thermodynamics, and Phase Transitions of Polymer Liquids and Alloys. Vol. 116, pp. 319-378.
Scranton, A. B., Rangarajan, B. and *Klier, J.*: Biomedical Applications of Polyelectrolytes. Vol. 122, pp. 1-54.
Sefton, M. V. and *Stevenson, W. T. K.*: Microencapsulation of Live Animal Cells Using Polycrylates. Vol.107, pp. 143-198.
Shamanin, V. V.: Bases of the Axiomatic Theory of Addition Polymerization. Vol. 112, pp. 135-180.
Sheiko, S. S.: Imaging of Polymers Using Scanning Force Microscopy: From Superstructures to Individual Molecules. Vol. 151, pp. 61-174.
Sherrington, D. C. see Cameron, N. R. , Vol. 126, pp. 163-214.
Sherrington, D. C. see Lin, J.: Vol. 111, pp. 177-220.
Sherrington, D. C. see Steinke, J.: Vol. 123, pp. 81-126.
Shibayama, M. see Tanaka, T.: Vol. 109, pp. 1-62.
Shiga, T.: Deformation and Viscoelastic Behavior of Polymer Gels in Electric Fields. Vol. 134, pp. 131-164.
Shim, H.-K., Jin, J.: Light-Emitting Characteristics of Conjugated Polymers. Vol. 158, pp. 191-241.
Shoda, S. see Kobayashi, S.: Vol. 121, pp. 1-30.
Siegel, R. A.: Hydrophobic Weak Polyelectrolyte Gels: Studies of Swelling Equilibria and Kinetics. Vol. 109, pp. 233-268.
Silvestre, F. see Calmon-Decriaud, A.: Vol. 207, pp. 207-226.
Sillion, B. see Mison, P.: Vol. 140, pp. 137-180.
Singh, R. P. see Sivaram, S.: Vol. 101, pp. 169-216.
Sinha Ray, S. see Biswas, M: Vol. 155, pp. 167-221.
Sivaram, S. and *Singh, R. P.*: Degradation and Stabilization of Ethylene-Propylene Copolymers and Their Blends: A Critical Review. Vol. 101, pp. 169-216.

Söderqvist Lindblad, M., Liu, Y., Albertsson, A.-C., Ranucci, E., Karlsson, S.: Polymer from Renewable Resources. Vol. 157, pp. 139–161
Starodybtzev, S. see Khokhlov, A.: Vol. 109, pp. 121-172.
Stegeman, G. I.: see Canva, M.: Vol. 158, pp. 87-121.
Steinke, J., Sherrington, D. C. and *Dunkin, I. R.*: Imprinting of Synthetic Polymers Using Molecular Templates. Vol. 123, pp. 81-126.
Stenzenberger, H. D.: Addition Polyimides. Vol. 117, pp. 165-220.
Stevenson, W. T. K. see Sefton, M. V.: Vol. 107, pp. 143-198.
Stridsberg, K. M., Ryner, M., Albertsson, A.-C.: Controlled Ring-Opening Polymerization: Polymers with Designed Macromoleculars Architecture. Vol. 157, pp. 27–51.
Suematsu, K.: Recent Progress of Gel Theory: Ring, Excluded Volume, and Dimension. Vol. 156, pp. 136-214.
Sumpter, B. G., Noid, D. W., Liang, G. L. and *Wunderlich, B.*: Atomistic Dynamics of Macromolecular Crystals. Vol. 116, pp. 27-72.
Sumpter, B. G. see Otaigbe, J.U.: Vol. 154, pp. 1-86.
Sugimoto, H. and *Inoue, S.*: Polymerization by Metalloporphyrin and Related Complexes. Vol. 146, pp. 39-120.
Suter, U. W. see Gusev, A. A.: Vol. 116, pp. 207-248.
Suter, U. W. see Leontidis, E.: Vol. 116, pp. 283-318.
Suter, U. W. see Rehahn, M.: Vol. 131/132, pp. 1-475.
Suter, U. W. see Baschnagel, J.: Vol. 152, p. 41-156.
Suzuki, A.: Phase Transition in Gels of Sub-Millimeter Size Induced by Interaction with Stimuli. Vol. 110, pp. 199-240.
Suzuki, A. and *Hirasa, O.*: An Approach to Artifical Muscle by Polymer Gels due to Micro-Phase Separation. Vol. 110, pp. 241-262.

Tagawa, S.: Radiation Effects on Ion Beams on Polymers. Vol. 105, pp. 99-116.
Tan, K. L. see Kang, E. T.: Vol. 106, pp. 135-190.
Tanaka, H. and *Shibayama, M.*: Phase Transition and Related Phenomena of Polymer Gels. Vol. 109, pp. 1-62.
Tanaka, T. see Penelle, J.: Vol. 102, pp. 73-104.
Tauer, K. see Guyot, A.: Vol. 111, pp. 43-66.
Teramoto, A. see Sato, T.: Vol. 126, pp. 85-162.
Terent'eva, J. P. and *Fridman, M. L.*: Compositions Based on Aminoresins. Vol. 101, pp. 29-64.
Theodorou, D. N. see Dodd, L. R.: Vol. 116, pp. 249-282.
Thomson, R. C., Wake, M. C., Yaszemski, M. J. and *Mikos, A. G.*: Biodegradable Polymer Scaffolds to Regenerate Organs. Vol. 122, pp. 245-274.
Tokita, M.: Friction Between Polymer Networks of Gels and Solvent. Vol. 110, pp. 27-48.
Tries, V. see Baschnagel, J:. Vol. 152, p. 41-156.
Tsuruta, T.: Contemporary Topics in Polymeric Materials for Biomedical Applications. Vol. 126, pp. 1-52.

Uyama, H. see Kobayashi, S.: Vol. 121, pp. 1-30.
Uyama, Y: Surface Modification of Polymers by Grafting. Vol. 137, pp. 1-40.

Varma, I. K. see Albertsson, A.-C.: Vol. 157, pp. 99-138.
Vasilevskaya, V. see Khokhlov, A.: Vol. 109, pp. 121-172.
Vaskova, V. see Hunkeler, D.: Vol.:112, pp. 115-134.
Verdugo, P.: Polymer Gel Phase Transition in Condensation-Decondensation of Secretory Products. Vol. 110, pp. 145-156.
Vettegren, V. I.: see Bronnikov, S. V.: Vol. 125, pp. 103-146.
Viovy, J.-L. and *Lesec, J.*: Separation of Macromolecules in Gels: Permeation Chromatography and Electrophoresis. Vol. 114, pp. 1-42.
Vlahos, C. see Hadjichristidis, N.: Vol. 142, pp. 71-128.

Volksen, W.: Condensation Polyimides: Synthesis, Solution Behavior, and Imidization Characteristics. Vol. 117, pp. 111-164.
Volksen, W. see Hedrick, J. L.: Vol. 141, pp. 1-44.
Volksen, W. see Hedrick, J. L.: Vol. 147, pp. 61-112.

Wake, M. C. see Thomson, R. C.: Vol. 122, pp. 245-274.
Wandrey C., Hernández-Barajas, J. and *Hunkeler, D.*: Diallyldimethylammonium Chloride and its Polymers. Vol. 145, pp. 123-182.
Wang, K. L. see Cussler, E. L.: Vol. 110, pp. 67-80.
Wang, S.-Q.: Molecular Transitions and Dynamics at Polymer/Wall Interfaces: Origins of Flow Instabilities and Wall Slip. Vol. 138, pp. 227-276.
Wang, T. G. see Prokop, A.: Vol. 136, pp.1-52; 53-74.
Whitesell, R. R. see Prokop, A.: Vol. 136, pp. 53-74.
Williams, R. J. J., Rozenberg, B. A., Pascault, J.-P.: Reaction Induced Phase Separation in Modified Thermosetting Polymers. Vol. 128, pp. 95-156.
Winter, H. H., Mours, M.: Rheology of Polymers Near Liquid-Solid Transitions. Vol. 134, pp. 165-234.
Wu, C.: Laser Light Scattering Characterization of Special Intractable Macromolecules in Solution. Vol 137, pp. 103-134.
Wunderlich, B. see Sumpter, B. G.: Vol. 116, pp. 27-72.

Xiang, M. see Jiang, M.: Vol. 146, pp. 121-194.
Xie, T. Y. see Hunkeler, D.: Vol. 112, pp. 115-134.
Xu, Z., Hadjichristidis, N., Fetters, L. J. and *Mays, J. W.*: Structure/Chain-Flexibility Relationships of Polymers. Vol. 120, pp. 1-50.

Yagci, Y. and *Endo, T.*: N-Benzyl and N-Alkoxy Pyridium Salts as Thermal and Photochemical Initiators for Cationic Polymerization. Vol. 127, pp. 59-86.
Yannas, I. V.: Tissue Regeneration Templates Based on Collagen-Glycosaminoglycan Copolymers. Vol. 122, pp. 219-244.
Yang, J. S. see Jo, W. H.: Vol. 156, pp. 1-52.
Yamaoka, H.: Polymer Materials for Fusion Reactors. Vol. 105, pp. 117-144.
Yasuda, H. and *Ihara, E.*: Rare Earth Metal-Initiated Living Polymerizations of Polar and Nonpolar Monomers. Vol. 133, pp. 53-102.
Yaszemski, M. J. see Thomson, R. C.: Vol. 122, pp. 245-274.
Yoo, T. see Quirk, R.P.: Vol. 153, pp. 67-162.
Yoon, D. Y. see Hedrick, J. L.: Vol. 141, pp. 1-44.
Yoshida, H. and *Ichikawa, T.*: Electron Spin Studies of Free Radicals in Irradiated Polymers. Vol. 105, pp. 3-36.

Zhou, H. see Jiang, M.: Vol. 146, pp. 121-194.
Zubov, V. P., Ivanov, A. E. and *Saburov, V. V.*: Polymer-Coated Adsorbents for the Separation of Biopolymers and Particles. Vol. 104, pp. 135-176.

Subject Index

Adriamycin 60
Agglomerates 1, 8, 9, 11, 35, 41
Agglomeration-deagglomeration 7
Aggregates 8, 16, 32
Albumin microspheres 92
Alginate 76
Amine oxide 68
Aminoalkyl 78
Anesthetics 62, 89
Animal data 134
Anti-inflammatory drugs 56
Antibiotic therapy 62
Anticancer drugs 60
Antigen delivery 119, 152
Antihypertensive drugs 62
Antitumor drugs 56, 59
Arrhenius temperature behavior 32

Beads, multiporous 75
Bending-twisting energy/modulus 31, 32
Bioavailability 122
Biocompatibility 154
Biopolyanhydrides 103
Blending 106
Block copolymers 68, 99
Blue dextran 79
Bonds (singly connected) 25–29
Bovine serum albumin 126
Bromoacyl bromide 78
Bupivacaine 89

Calcium chloride 124
ε-Caprolactone 58
Carbon black 1–13, 16, 20–23, 28–41
– –, strain 4, 5
Carbopol 934 93
Carboxymethyl cellulose 93
Carboxymethyl chitin 86
Carboxyvinyl polymer 93
Carrageenan 66
Castor oil 105

CCA-clusters 30, 31, 34, 37
Cellulose, carboxymethyl 93
–, crystalline 95
–, hydroxypropyl 93
Cellulose acetate phthalate 93
Cellulose ethers 63
Cellulose sulfate 124
Chemotheraputic carrier, adriamycin 76
Chitosan 60, 66, 75, 94, 126
Chitosan glutamate 124
Chitosan-amine oxide gel 68
Chitosan-PEO 60
Chitosans, N-acyl 92
Chloramphenicol 68
Chlorhexidini acetas 67
Chlorothiazide 92
Cimetidine 67, 68
Citric acid 105
Cluster-cluster aggregation model, filler network 30
Cole-Cole 10, 14, 15, 20
Compatibility 134
Complex modulus 8, 17, 19, 40
Complexation 158
Connectivity exponent 12
Contact breakage 9
Controlled drug release 45
Core-shell morphology, nanoparticle 137
Coulter DELSA 440SX 132
CPA-behavior 16
CPM 88
Cytochrome C 125
Cytolytic activity 99
Cytotoxic activity 99

Delivery 45, 58, 64, 80, 119, 152, 168
Delivery, antigen 119, 152
–, controlled 168
DFS 88
Diacids, aliphatic 103
Dibucaine 89

Dilactide microspheres 88
Diltiazem 64
1,5-Diozepan-2-one 88
Diphtheria toxoid 126
Discomes 72
Disintegration agent 96
Dissipated energy 23, 41
Divinylbenzene 65
Drug delivery 119
– –, intravenous 58
Drug delivery systems 45
Drug loading 161
Drug release, controlled 45
Drug targeting 59
Dynamic modulus 1, 3, 5, 13, 27

Elastic constants 26
Elastic modulus 1, 31, 32, 34–36, 40
Elasticity exponent 38
Elastomers 1
Electrophoretic mobility 132
Emulsifier 99
Emulsion polymerization 52
Energy loss density 3
Entrapment efficiency 131, 137, 161
Enzyme carriers 60
Ester linkages 63
Ethyl cellulose 93, 96

FAD 103
Failure strain 24, 28
Fibroblast growth factor-2 125
Filler network 1, 4–13, 16, 19, 22–41
– –, network junction model 22
Filler structures, fractal 1
Films/membranes 92
Flurbiprofen 72
Force constant 26, 27
Formaldehyde 92
Fractal dimension 10, 12, 16, 31, 32
Fractal filler structures 1
Freeze-drying 133, 148
Frictional models 9
5-FU 86

β-Galactosidase 64
Gelatin 63
Gelatin films, cross-linked 93
Gelatin microspheres 87
Gelation, ionic 60
Gellan 124
Gellan gum 81
Gene therapy 58, 100
Glass transition 32, 33
Glutamate microspheres 90

Glutaraldehyde 67
Glycine 88
Glycopyranose 59

Havriliak-Negami function 38
Health risks, hyrdophobic toxic organics 136
Hemoglobin 79
Heparin 124
Hexyl cyanoacrylate 52
Hilbert transform 14
Hormone therapy 62
Hydrodynamic reinforcement 16, 40
Hydrogels 45, 61
–, bioerodable 62
–, superporous 65, 66
Hydrophilic drugs 119
Hydroxypropyl cellulose 93
Hydroxypropyl methylcellulose 95
Hydroxystearic acid 105
Hysteresis 1, 16–18, 22, 24, 40, 41
Hysteresis cycle 18

Ibuprofen 61
IdUrd 82
Immobilised rubber 1, 34–37, 41
Immunizations 135
Immunoassays 135
Indomethacin 64, 97
INH 88
Insulin loading 59
Interaggregate contacts 17, 20
Interface slippage 23, 41
Intraperitoneal application 64

Junction gap width 23
Junction points 22
Junction shape factor 24

Ketoprofen 61
Kramers-Kronig relation 14
Kraus model 9, 17

Lactose 96
Laser Doppler anemometry 132
Levamisole 69
Lidocaine 89
Linear viscoelastic rubber 2
Linear viscoelastic behavior 17
Links-nodes-blobs model, filler network 1, 24, 25
Lipids 86
Liposomes, mucoadhesive 98
Lipospheres 89

Subject Index

Loading efficiency 131, 137, 161
Loss modulus 2–16
Lycoprotectant 101

Macromolecular drugs 76
Macromolecules, amphiphilic 51
Maleic anhydride 93
Methacrylic acid 65
Methotrxate-O-carboxymethylate chitosan 60
Methoxyflurane 89
Methoxypoly(ethylene glycol) 100
Methyl vinyl ether 93
Microcapsules 75
Microgels 4, 30, 36, 38
Microparticles 45
Microscopic complex modulus 17
Microsphere gel 81
Microspheres 75
Monolithic gels 70
Monomethoxymonoamine 58
Mucoadhesive liposomes 98

Nanocapsule 51
Nanoemulsions 56
Nanoparticles 45, 50
–, assembly/structure 136
–, hydrogel-based colloidal system 119
–, size/charge, batch/continuous processes 119, 123, 139
Nanosphere 51
Nasal application 64
Network junction model, filler network 2, 22
Niosomes 45, 70
–, polyhedral 74

Ocular drug delivery 64, 80
Offsets 6, 7
Oligonucleotides 59
Ophthalmic application 64, 80
Oral cavity 95
Ovalbumine 124

Paclitaxel 73
Particle size analysis, colloidal systems 54
Patches, polymeric 94
Payne effect 1, 11–13, 28, 38, 40, 41
PBG-PEO 59
PCS 132
Pectin 95
–, low-ethoxylated 126
Pentasodium tripolyphosphate 124
PEO 56, 126
Peptide transport 59

Peptides 84
Percolation 1, 12, 24–29, 37–41
Permeability 148
Permeability control 119
pH sensitivity 61
Phospholipids 62, 90
Photon correlation spectrometry 132
PLA-PEO 55
Plasmid complexes 101
PLG-PPO-PEO 56
Pluronic F-68 124
Pluronic F-127 64
PMCG 124
Poloxamer 99
Poly(adipic anhydride) 80, 106
Poly(β-benzyl-L-aspartate) 56
Poly(ethylene glycol) coated nanospheres 58
Poly(isobutyl cyanoacrylate) 59
Poly(lactic acid) 56
Poly(D,L-lactide-co-glycolide) 82
Poly(methacrylic acid) 65
Poly(methylene-co-guanidine) hydrochloride 124
Poly(ortho esters) 62
Poly(trimethyl carbonate)-poly(adipic anhydride) 106
Poly(vinyl acetate) 93
Polyacrylamide 63
Polyallylamine 78
Polyamides 106
Polyamines 78, 124
Polyanhydrides 62, 90
–, aliphatic 103
Polycation microencapsulation 76
Polycondensation 105
Polydextran aldehyde 125
Polyelectrolyte 67, 91, 157
Polyethylene glycol 126
Polyethylene oxide-polypropylene oxide 126
Polyethyleneimine 78
Polylactide-co-glycolide 55
Poly-L-lysine 84
Polymer precipitation 53
Polymer tablets 95
Polypeptides, γ-benzylglutamic acid 101
Polyphosphazene derivatives 58
Polysaccharides 86
Polystyrene 65
Polyvinylamine 125
Potato starch 96
Pranoprofen 61
Primary aggregates 24
Product stability 146

Propanolol-hydrochloride 82, 94
Protease, immobilized 76
Protein cross-linking 163
Protein delivery 119
Protein drug delivery 61
Protein loading 60
Proteins, small size 164

Rate equilibrium 37
Release 162 ff.
Reticuloendothelial system 59
Ricinoleic acid 105
Ricinoleics 104
Rigidy condition 69

SA 103
Salting-out, drug encapsulation 54
Scaling invariance 34
Self-similar 31
Shape factor 23
Shear stress/modulus 2, 3
Silica 4, 22, 35
Silver sulfadiazene 68
Sinusoidal strain 3
Site-specific delivery 61
Slip-links 1, 24, 41
Sodium alginate 79
Sodium hexametaphosphate 125, 126
Solid dosage 63
Solid lipid nanoparticles 61
Solvent displacement/nanoprecipitation 53
Solvent extraction-evaporation 53
Space filling configuration/condition 37
Spacer groups 88
Spermine tetrahydrochloride 124
Stabilization, steric 160
Star-shaped (co)polymers 65
Starch 96
Steric stabilization 160
Storage modulus 1–8, 12, 13, 19, 28, 30, 32–39
Stored energy 11, 23, 31, 39
Strain 2
–, energy loss density 3
Strain energy 8, 26

Strain sweeps 20
Styrene-butadiene rubber 20
Subcutaneous application 64
Sulfadiazine 105
Surface charge 54
Surface properties 160
System stoichiometry 145

Tablets 45
Tamarind seeds 64
Tectal application 64
TEISI model 16
Terbutaline sulfate 95
Tetanus toxoid 125
– – antigen 126
Tetracaine 61, 89
Theophylline 63, 97
Thixotropic change 6
Thy-HCl 88
Timolol maleate 81
Titration, nanoparticle production 123
Topical application 64
Transdermal application 45
Transdermal devices 94
TRH microspheres 90
Triglyceride liposphers 89
Trimethylene carbonate 88
Tripolyphosphate 126
Tris(6-isocyanatohexyl) isocyanurate 62

Urethane/allophanate 62

Vaccine delivery 152
Vaccine vehicles 164
van der Waals interaction 8
Viscoelasticity 1
VTG model 17

Water-soluble polymers 119

Xyloglucan gels 64

Young's modulus 23

Zener model 11, 12

You are one **click** *away from a* **world of chemistry** *information!*

Come and visit Springer's
Chemistry Online Library

Books
- Search the Springer website catalogue
- Subscribe to our free alerting service for new books
- Look through the book series profiles

You want to order? Email to: orders@springer.de

Journals
- Get abstracts, ToC´s free of charge to everyone
- Use our powerful search engine LINK Search
- Subscribe to our free alerting service LINK *Alert*
- Read full-text articles (available only to subscribers of the paper version of a journal)

You want to subscribe? Email to: subscriptions@springer.de

Electronic Media
- Get more information on our software and CD-ROMs

You have a question on an electronic product? Email to: helpdesk-em@springer.de

Bookmark now:

http://www.springer.de/geosci/

 Springer

Springer · Customer Service
Haberstr. 7 · D-69126 Heidelberg, Germany
Tel: +49 6221 345-217/218 · Fax: +49 6221 345-229
d&p · 006756_001x_1c

Printing (Computer to Film): Saladruck, Berlin
Binding: Stürtz AG, Würzburg